Geotechnical Engineering Design

Geotechnical Engineering Design

Editor

Petre Kozel

scitus
academics

Geotechnical Engineering Design
Edited by **Petre Kozel**

Printed in 2017

ISBN: 978-1-68117-141-8
Library of Congress Control Number: 2015950361

Table of Content

Preface

Geotechnical engineering is the branch of civil engineering concerned with the engineering behavior of earth materials. Geotechnical engineering is important in civil engineering, but also has applications in military, mining, petroleum and other engineering disciplines that are concerned with construction occurring on the surface or within the ground. Geotechnical engineering uses principles of soil mechanics and rock mechanics to investigate subsurface conditions and materials; determine the relevant physical/mechanical and chemical properties of these materials; evaluate stability of natural slopes and man-made soil deposits; assess risks posed by site conditions; design earthworks and structure foundations; and monitor site conditions, earthwork and foundation construction.

This book presents the fundamental design principles and approaches in geotechnical engineering. including an introduction to engineering geology, subsurface explorations, shallow and deep foundations, slope stability analyses and remediation, filters and drains, Earth retaining structures, geosynthetics, and bask seismic evaluations of slope stability, lateral earth pressures, and liquefaction. This book applies the principles of soil mechanics, earthquake and focuses on the design methodologies in geotechnical engineering.

CHAPTER 1

A Cognitive Look at Geotechnical Earthquake Engineering: Understanding the Multidimensionality of the Phenomena

Silvia Garcia

Geotechnical Department, Institute of Engineering, National University of Mexico, Mexico

INTRODUCTION

Not even windstorm, earth-tremor, or rush of water is a catastrophe.
A catastrophe is known by its works; that is, to say, by the occurrence of disaster.
So long as the ship rides out the storm, so long as the city resists the earth-shocks, so long as the levees hold, there is no disaster.
It is the collapse of the cultural protections that constitutes the proper disaster.

(Carr, 1932)

Essentially, disasters are human-made. For a catastrophic event, whether precipitated by natural phenomena or human activities, assumes the state of a disaster when the community or society affected fails to cope. Earthquake hazards themselves do not necessarily lead to disasters, however intense, inevitable or unpredictable, translate to disasters only to the extent that the population is unprepared to respond, unable to deal with, and, consequently, severely affected. Seismic disasters could, in fact, be reduced if not prevented. With today's advancements in science and technology, including early warning and forecasting of the natural phenomena, together with innovative approaches and strategies for enhancing local capacities, the impact of earthquake hazards somehow could be predicted and mitigated, its detrimental effects on populations reduced, and the communities adequately protected.

After each major earthquake, it has been concluded that the experienced ground motions were not expected and soil behavior and soil-structure interaction were not properly predicted. Failures, associated to inadequate design/construction and to lack of phenomena comprehension, obligate further code reinforcement and research. This scenario will be repeated after each earthquake. To overcome this issue, *Earthquake Engineering* should change its views on the present methodologies and techniques toward more scientific, doable, affordable, robust and adaptable solutions.

A competent modeling of engineering systems, when they are affected by seismic activity, poses many difficult challenges. Any representation designed for reasoning about models of such systems has to be flexible enough to handle various degrees of complexity and uncertainty, and at the same time be sufficiently powerful to deal with situations in which the input signal may or may not be controllable. Mathematically-based models are developed using scientific theories and concepts that just apply to particular conditions. Thus, the core of the model comes from assumptions that for complex systems usually lead to simplifications (perhaps oversimplifications) of the problem phenomena. It is fair to argue that the representativeness of a particular theoretical model largely depends on the degree of comprehension the developer has on the behavior of the actual engineering problem. Predicting natural-phenomena characteristics like those of earthquakes, and thereupon their potential effects at particular sites, certainly belong to a class of problems we do not fully understand. Accordingly, analytical modeling often becomes the bottleneck in the development of more accurate procedures. As a consequence, a strong demand for advanced modeling an identification schemes arises.

Cognitive Computing CC technologies have provided us with a unique opportunity to establish coherent seismic analysis environments in which uncertainty and partial data-knowledge are systematically handled. By seamlessly combining learning, adaptation, evolution, and fuzziness, CC complements current engineering approaches allowing us develop a more comprehensive and unified framework to the effective management of earthquake phenomena. Each CC algorithm has well-defined labels and could usually be identified with specific scientific communities. Lately, as we improved our understanding of these algorithms' strengths and weaknesses, we began to leverage their best features and developed hybrid algorithms that indicate a new trend of co-existence and integration between many scientific communities to solve a specific task.

In this chapter geotechnical aspects of earthquake engineering under a cognitive examination are covered. Geotechnical earthquake engineering, an area that deals with the design and construction of projects in order to resist the effect of earthquakes, requires an understanding of geology, seismology and earthquake engineering. Furthermore, practice of geotechnical earthquake engineering also requires consideration of social, ECONOMIC AND POLITICAL FACTORS. VIA THE DEVELOPMENT OF COGNITIVE interpretations of selected topics: i) spatial variation of soil dynamic properties, ii) attenuation laws for rock sites (seismic input), iii) generation of artificial-motion time histories, iv) effects of local site conditions (site effects), and iv) evaluation of liquefaction susceptibility, CC techniques (Neural Networks NNs, Fuzzy Logic FL and Genetic Algorithms GAs) are presented as appealing alternatives for integrated data-driven and theoretical procedures to generate reliable seismic models.

GEOTECHNICAL EARTHQUAKE HAZARDS

The author is well aware that standards for geotechnical seismic design are under development worldwide. While there is no need to "reinvent the wheel" there is a requirement to adapt such initiatives to fit the emerging safety philosophy and demands. This investigation also strongly endorses the view that "guidelines" are far more desirable than "codes" or "standards" disseminated all over seismic regions. Flexibility in approach is a key ingredient of geotechnical engineering and the cognitive technology in this area is rapidly advancing.The science and practice of geotechnical earthquake engineering is far from mature and need to be expanded and revised periodically in coming years. It is important that readers and users of the computational models presented here familiarize themselves with the latest advances and amend the recommendations herein appropriately.

This document is not intended to be a detailed treatise of latest research in geotechnical earthquake engineering, but to provide sound guidelines to support rational cognitive approaches. While every effort has been made to make the material useful in a wider range of applications, applicability of the material is a matter for the user to judge. The main aim of this guidance document is to promote consistency of cognitive approach to everyday situations and, thus, improve geotechnical-earthquake aspects of the performance of the built safe-environment.

A "Soft" Interpretation of Ground Motions

After a sudden rupture of the earth's crust (caused by accumulating stresses, elastic strain-energy) a certain amount of energy radiates from the rupture as seismic waves. These waves are attenuated, refracted, and reflected as they travel through the earth, eventually reaching the surface where they cause ground shaking. The principal geotechnical hazards associated with this event are fault rupture, ground shaking, liquefaction and lateral spreading, and landsliding. Ground shaking is one of the principal seismic hazards that causes extensive damage to the built environment and failure of engineering systems over large areas. Earthquake loads and their effects on structures are directly related to the intensity and duration of ground shaking. Similarly, the level of ground deformation, damage to earth structures and ground failures are closely related to the severity of ground shaking.

In engineering evaluations, three characteristics of ground shaking are typically considered: i) the amplitude, ii) frequency content and iii) significant duration of shaking (time over which the ground motion has relatively significant amplitudes).These characteristics of the ground motion at a given site are affected by numerous complex factors such as the source mechanism, earthquake magnitude, rupture directivity, propagation path of seismic waves, source distance and effects of local soil conditions. There are many unknowns and uncertainties associated with these issues which in turn result in significant uncertainties regarding the characteristics of the ground motion and earthquake loads.

If the random nature of response to earthquakes (aleatory uncertainty) cannot be avoided [1,2], it is our limited knowledge about the patterns between seismic events and their manifestations -ground motions- at a site (epistemic uncertainty) that must be improved thorough more scientific seismic analyses. A strategic factor in seismic hazard analysis is the ground motion model or attenuation relation. These attenuation relationships has been developed based on magnitude, distance and site category, however, there is a tendency to incorporate other parameters, which are now known to be significant, as the tectonic environment, style of faulting and the effects of topography, deep basin edges and rupture directivity. These distinctions are recognized in North America, Japan and New Zealand [3-6], but ignored in most other regions of the world [7]. Despite recorded data suggest that ground motions depend, in a significant way, on these aspects, these inclusions did not have had a remarkable effect on the predictions confidence and the geotechnical earthquake engineer prefers the basic and clear-cut approximations on those that demand a *blind* use of coefficients or an intricate determination of soil/fault conditions.

A key practice in current aseismic design is to develop design spectrum compatible time histories. This development entails the modification of a time history so that its response spectrum matches within a prescribed tolerance level, the target design spectrum. In such matching it is important to retain the phase characteristics of the selected ground motion time history. Many of the techniques used to develop compatible motions do not retain the phase [8]. The response spectrum alone does not adequately characterize specific-fault ground motion. Near-fault ground motions must be characterized by a long period pulse of strong motion of a fairly brief duration rather than the stochastic process of long duration that characterizes more distant ground motions. Spectrum compatible with these specific motions will not have these characteristics unless the basic motion being modified to ensure compatibility has these effects included. Spectral compatible motions could match the entire spectrum but the problem arises on finding a "real" earthquake time series that match the specific nature of ground motion. For nonlinear analysis of structures, spectrum compatible motions should also correspond to the particular energy input [9], for this reason, designers should be cautious about using spectrum compatible motions when estimating the displacements of embankment dams and earth structures under strong shaking, if the acceptable performance of these structures is specified by criteria based on tolerable displacements.

Another important seismic phenomenon is the liquefaction. Liquefaction is associated with significant loss of stiffness and strength in the shaken soil and consequent large ground deformation. Particularly damaging for engineering structures are cyclic ground movements during the period of shaking and excessive residual deformations such as settlements of the ground and lateral spreads. Ground surface disruption including surface cracking, dislocation, ground distortion, slumping and permanent deformations, large settlements and lateral spreads are commonly observed at liquefied sites. In sloping ground and backfills behind retaining structures in waterfront areas, liquefaction often results in large permanent ground displacements in the down-slope direction or towards waterways (lateral spreads). Dams, embankments and sloping ground near riverbanks where certain shear strength is required for stability under gravity loads are particularly prone to such failures. Clay soils may also suffer some loss of strength during shaking but are not subject to boils and other "classic" liquefaction phenomena. For intermediate soils, the transition from "sand like" to "clay-like" behavior depends primarily on whether the soil is a matrix of coarse grains with fines contained within the pores or a matrix of plastic fines with coarse grained "filler". Recent papers by Boulanger and Idriss [10, 11] are helpful in clarifying issues surrounding the liquefaction and strain softening of different soil types during strong ground shaking. Engineering judgment based on good quality investigations and data interpretation should be used for classifying such soils as liquefiable or non-liquefiable.

Procedures for evaluating liquefaction, potential and induced lateral spread, have been studied by many engineering committees around the world. The objective has been to review research and field experience on liquefaction and recommended standards for practice. Youd and Idriss [12] findings and the liquefaction-resistance chart proposed by Seed et al. [13] in 1985, stay as standards for practice. They have been slightly modified to adjust new registered input-output conditions and there is a strong tendency to recommend i) the adoption of the cone penetration test CPT, standard penetration test SPT or the shear wave velocities for describing the *in situ* soil conditions [14] and ii) the modification of magnitude factors used to convert the critical stress ratios from the liquefaction assessment charts (usually developed for M7:5) to those appropriate for earthquakes of diverse magnitudes [12, 15].

COGNITIVE COMPUTING

Cognitive Computing CC as a discipline in a narrow sense, is an application of computers to solve a given computational problem by imperative instructions; while in a broad sense, it is a process to implement the instructive intelligence by a system that transfers a set of given information or instructions into expected behaviors. According to theories of cognitive informatics [16-18], computing technologies and systems may be classified into the categories of imperative, autonomic, and cognitive from the bottom up. Imperative computing is a traditional and passive technology based on stored-program controlled behaviors for data processing [19-24]. An autonomic computing is goal-driven and self-decision-driven technologies that do not rely on instructive and procedural information [25-28]. Cognitive computing is more intelligent technologies beyond imperative and autonomic computing, which embodies major natural intelligence behaviors of the brain such as thinking, inference, learning, and perceptions.

Cognitive computing is an emerging paradigm of intelligent computing methodologies and systems, which implements computational intelligence by autonomous inferences and perceptions mimicking the mechanisms of the brain. This section presents a brief description on the theoretical framework and architectural techniques of cognitive computing beyond conventional imperative and autonomic computing technologies. Cognitive models are explored on the basis of the latest advances in applying computational intelligence. These applications of cognitive computing are described from the aspects of cognitive search engines, which demonstrate how machine and computational intelligence technologies can drive us toward autonomous knowledge processing.

Computational Intelligence: Soft Computing Technologies

The *computational intelligence* is a synergistic integration of essentially three computing paradigms, viz. neural networks, fuzzy logic and evolutionary computation entailing probabilistic reasoning (belief networks, genetic algorithms and chaotic systems) [29]. This synergism provides a framework for flexible information processing applications designed to operate in the real world and is commonly called *Soft Computing SC* [30]. Soft computing technologies are robust by design, and operate by trading off precision for tractability. Since they can handle uncertainty with ease, they conform better to real world situations and provide lower cost solutions.

The three components of soft computing differ from one another in more than one way. Neural networks operate in a numeric framework, and are well known for their learning and generalization capabilities. Fuzzy systems [31] operate in a linguistic framework, and their strength lies in their capability to handle linguistic information and perform approximate reasoning. The evolutionary computation techniques provide powerful search and optimization methodologies. All the three facets of soft computing differ from one another in their time scales of operation and in the extent to which they embed *a priori* knowledge.

Figure 1 shows a general structure of Soft Computing technology. The following main components of SC are known by now: fuzzy logic FL, neural networks NN, probabilistic reasoning PR, genetic algorithms GA, and chaos theory ChT (Figure 1). In SC FL is mainly concerned with imprecision and approximate reasoning, NN with learning, PR with uncertainty and propagation of belief, GA with global optimization and search and ChT with nonlinear dynamics. Each of these computational paradigms (emerging reasoning technologies) provides us with complementary reasoning and searching methods to solve complex, real-world problems. In large scope, FL, NN, PR, and GA are complementary rather that competitive [32-34]. The interrelations between the components of SC, shown in Figure 1, make the theoretical foundation of Hybrid Intelligent Systems. As noted by L. Zadeh: "... the term hybrid intelligent systems is gaining currency as a descriptor of systems in which FL, NC, and PR are used in combination. In my view, hybrid intelligent systems are the wave of the future" [35]. The use of Hybrid

Intelligent Systems are leading to the development of numerous manufacturing system, multimedia system, intelligent robots, trading systems, which exhibits a high level of MIQ (machine intelligence quotient).

Comparative Characteristics of SC Tools

The constituents of SC can be used independently (fuzzy computing, neural computing, evolutionary computing etc.), and more often in combination [36, 37, 38- 40, 41]. Based on independent use of the constituents of Soft Computing, fuzzy technology, neural technology, chaos technology and others have been recently applied as emerging technologies to both industrial and non-industrial areas.

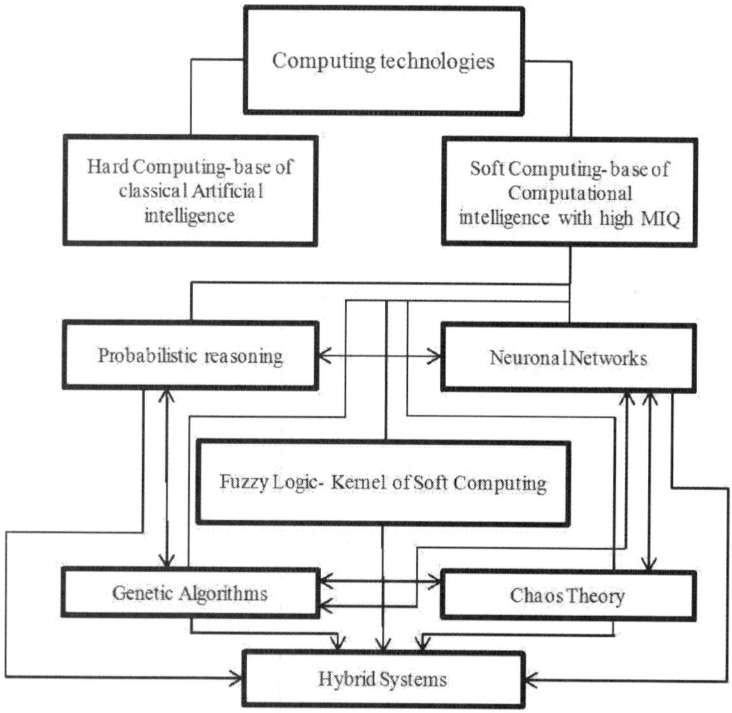

Figure 1. Soft Computing Components

Fuzzy logic is the leading constituent of Soft Computing. In Soft Computing, fuzzy logic plays a unique role. FL serves to provide a methodology for computing [36]. It has been successfully applied to many industrial spheres, robotics, complex decision making and diagnosis, data compression, and many other areas. To design a system processor for handling knowledge represented in a linguistic or uncertain numerical form we need a fuzzy model of the system. Fuzzy sets can be used as a universal approximator, which is very important for modeling unknown objects. If an operator cannot tell linguistically what kind of action he or she takes in a specific situation, then it is quite useful to model his/her control actions using numerical data. However, fuzzy logic in its so called *pure form* is not always useful for easily constructing intelligent systems. For example, when a designer does not have sufficient prior information (knowledge) about the system, development of acceptable fuzzy rule base becomes impossible. As the complexity of the system increases, it becomes difficult to specify a correct set of rules and membership functions for describing adequately the behavior of the system. Fuzzy systems also have the disadvantage of not being able to extract additional knowledge from the experience and correcting the fuzzy rules for improving the performance of the system.

Another important component of Soft Computing is neural networks. Neural networks NN viewed as parallel computational models, are parallel fine-grained implementation of non-linear static or dynamic systems. A very important feature of these networks is their adaptive nature, where "learning by example" replaces traditional "programming" in problems solving. Another key feature is the intrinsic parallelism that allows fast computations. Neural networks are viable computational models for a wide variety of problems including pattern classification, speech synthesis and recognition, curve fitting, approximation capability, image data compression, associative memory, and modeling and control of non-linear unknown systems [42, 43]. NN are favorably distinguished for efficiency of their computations and hardware implementations. Another advantage of NN is generalization ability, which is the ability to classify correctly new patterns. A significant disadvantage of NN is their poor interpretability. One of the main criticisms addressed to neural networks concerns their black box nature [35].

Evolutionary Computing EC is a revolutionary approach to optimization. One part of EC—genetic algorithms—are algorithms for global optimization. Genetic algorithms GAs are based on the mechanisms of natural selection and genetics [44]. One advantage of genetic algorithms is that they effectively implement parallel multi-criteria search. The mechanism of genetic algorithms is simple. Simplicity of operations and powerful computational effect are the two main advantages of genetic algorithms. The disadvantages are the problem of convergence and the absence of strong theoretical foundation. The requirement of coding the domain of the real variables' into bit strings also seems to be a drawback of genetic algorithms. It should be also noted that the computational speed of genetic algorithms is low.

Because in this investigation PR and ChT are not exploited, they are not going to be explained. For the interested reader [41] is recommended. Table 1 presents the comparative characteristics of the components of Soft Computing. For each component of Soft Computing there is a specific class of problems, where the use of other components is inadequate.

Intelligent Combinations of SC

As it was shown above, the components of SC complement each other, rather than compete. It becomes clear that FL, NC and GA are more effective when used in combinations. Lack of interpretability of neural networks and poor learning capability of fuzzy systems are similar problems that limit the application of these tools. Neurofuzzy systems are hybrid systems which try to solve this problem by combining the learning capability of connectionist models with the interpretability property of fuzzy systems. As it was noted above, in case of dynamic work environment, the automatic knowledge base correction in fuzzy systems becomes necessary. On the other hand, artificial neural networks are successfully used in problems connected to knowledge acquisition using learning by examples with the required degree of precision.

Incorporating neural networks in fuzzy systems for fuzzification, construction of fuzzy rules, optimization and adaptation of fuzzy knowledge base and implementation of fuzzy reasoning is the essence of the Neurofuzzy approach.

Table 1. Central characteristics of Soft Computing technologies

	Fuzzy Sets	**Artificial Neural Networks**	**Evolutionary Computing, GA**	**Probabilistic Reasoning**	**Chaotic computing**
Weaknesses	•Knowledge acquisition •Learning	•Black Box interpretability	•Coding •Computational speed	•Limitation of the axioms of Probability Theory •Lack of complete knowledge •Copmputational complexity	•Computational complexity •Chaos identification complexity
Strengths	•Interpretability •Transparency •Plausibility •Graduality •Modeling •Reasoning •Tolerance to imprecision	•Learning •Adaptation •Fault tolerance •Curve fiting •Generalization ability •Approximation ability	Computational efficiency •Global optimization	•Rigorous framework •Well understanding	•Nonlinear dynamics simulation •Discovering chaos in observed data (with noise) •Determinig the predictability •Prediction startegies formulation

The combination of genetic algorithms with neural networks yields promising results as well. It is known that one of main problems in development of artificial neural systems is selection of a suitable learning method for tuning the parameters of a neural network (weights, thresholds, and structure). The most known algorithm is the "error back propagation" algorithm. Unfortunately, there are some difficulties with "back propagation". First, the effectiveness of the learning considerably depends on initial set of weights, which are generated randomly. Second, the "back propagation", like any other gradient-based method, does not avoid local minima. Third, if the learning rate is too slow, it requires too much time to find the solution. If, on the other hand, the learning rate is too high it can generate oscillations around the desired point in the weight space. Fourth, "back propagation" requires the activation functions to be differentiable. This condition does not hold for many types of neural networks. Genetic algorithms used for solving many optimization problems when the "strong" methods fail to find appropriate solution, can be successfully applied for learning neural networks, because they are free of the above drawbacks.

The models of artificial neurons, which use linear, threshold, sigmoidal and other transfer functions, are effective for neural computing. However, it should be noted that such models are very simplified. For example, reaction of a biological axon is chaotic even if the input is periodical. In this aspect the more adequate model of neurons seems to be chaotic. Model of a chaotic neuron can be used as an element of chaotic neural networks. The more adequate results can be obtained if using fuzzy chaotic neural networks, which are closer to biological computation. Fuzzy systems with If-Then rules can model non-linear dynamic systems and capture chaotic attractors easily and accurately. Combination of Fuzzy Logic and Chaos Theory gives us useful tool for building system's chaotic behavior into rule structure. Identification of chaos allows us to determine predicting strategies. If we use a Neural Network Predictor for predicting the system's behavior, the parameters of the strange attractor (in particular fractal dimension) tell us how much data are necessary to train the neural network. The combination of Neurocomputing and Chaotic computing technologies can be very helpful for prediction and control.

The cooperation between these formalisms gives a useful tool for modeling and reasoning under uncertainty in complicated real-world problems. Such cooperation is of particular importance for constructing perception-based intelligent information systems. We hope that the mentioned intelligent combinations will develop further, and the new ones will be proposed. These SC paradigms will form the basis for creation and development of Computational Intelligence.

COGNITIVE MODELS OF GROUND MOTIONS

The existence of numerous databases in the field of civil engineering, and in particular in the field of geotechnical earthquake, has opened new research lines through the introduction of analysis based on soft computing. Three methods are mainly applied in this emerging field: the ones based on the Neural Networks NN, the ones created using Fuzzy Sets FS theory and the ones developed from the Evolutionary Computation [45].

The SC hybrids used in this investigation are directed to tasks of prediction (classification and/or regression). The central objective is obtaining numerical and/or categorical values that mimic input-output conditions from experimentation and in situ measurements and then, through the recorded data and accumulated experience, predict future behaviors. The examples presented herein have been developed by an engineering committee that works for generating useful guidance to geotechnical practitioners with geotechnical seismic design. This effort could help to minimize the perceived significant and undesirable variability within geotechnical earthquake practice. Some urgency in producing the alternative guidelines was seen, after the most recent earthquakes disasters, as being necessary with a desire to avoid a long and protracted process. To this end, a two stage approach was suggested with the first stage being a cognitive interpretation of well-known procedures with appropriate factors for geotechnical design, and a posterior step identifying the relevant philosophy for a new geotechnical seismic design.

Spatial Variation of Soil Dynamic Properties

The spatial variability of subsoil properties constitutes a major challenge in both the design and construction phases of most geo-engineering projects. Subsoil investigation is an imperative step in any civil engineering project. The purpose of an exploratory investigation is to infer accurate information about actual soil and rock conditions at the site. Soil exploration, testing, evaluation, and field observation are well-established and routine procedures that, if carried out conscientiously, will invariably lead to good engineering design and construction. It is impossible to determine the optimum spacing of borings before an investigation begins because the spacing depends not only on type of structure but also on uniformity or regularity of encountered soil deposits. Even the most detail soil maps are not efficient enough for predicting a specific soil property because it changes from place to place, even for the same soil type. Consequently interpolation techniques have been extensively exploited. The most commonly used methods are kriging and co-kriging but for better estimations they require a great number of measurements available for each soil type, what is generally impossible.

Based on the high cost of collecting soil attribute data at many locations across landscape, new interpolation methods must be tested in order to improve the estimation of soil properties. The integration of GIS and Soft Computing SC offers a potential mechanism to lower the cost of analysis of geotechnical information by reducing the amount of time spent understanding data. Applying GIS to large sites, where historical data can be organized to develop multiple databases for analytical and stratigraphic interpretation, originates the establishment of spatial/chronological efficient methodologies for interpreting properties (soil exploration) and behaviors (in situ measured). GIS-SC modeling/simulation of natural systems represents a new methodology for building predictive models, in this investigation NN and GAs, nonparametric cognitive methods, are used to analyze physical, mechanical and geometrical parameters in a geographical context. This kind of spatial analysis can handle uncertain, vague and incomplete/redundant data when modeling intricate relationships between multiple variables. This means that a NN has not constraints about the spacing (minimum distance) between the drill holes used for building (training) the SC model. The NNs-GAs acts as computerized architectures that can approximate nonlinear functions of several variables, this scheme represent the relations between the spatial patterns of the stratigraphy without restrictive assumptions or excessive geometrical and physical simplifications.

The geotechnical data requirements (geo-referenced properties) for an easy integration of the SC technologies are explained through an application example: a geo-referenced three-dimensional model of the soils underlying Mexico City. The classification/prediction criterion for this very complex urban area is established according to two variables: the cone penetration resistance qc (mechanical property) and the shear wave velocity Vs (dynamic property). The expected result is a 3D-model of the soils underlying the city area that would eventually be improved for a more complex and comprehensive model adding others mechanical, physical or geometrical geo-referenced parameters.

Cone-tip penetration resistances and shear wave velocities have been measured along 16 bore holes spreaded throughout the clay deposits of Mexico City (Figure 2). This information was used as the set of examples inputs (latitude, longitude and depth) → output (qc /Vs). The analysis was carried out in an approximate area of 125 km2 of Mexico City downtown. It is important to point out that 20% of these patterns (sample points and complete variables information) are not used in the training stage; they will be presented for testing the generalization capabilities of the closed system components (once the training is stopped).

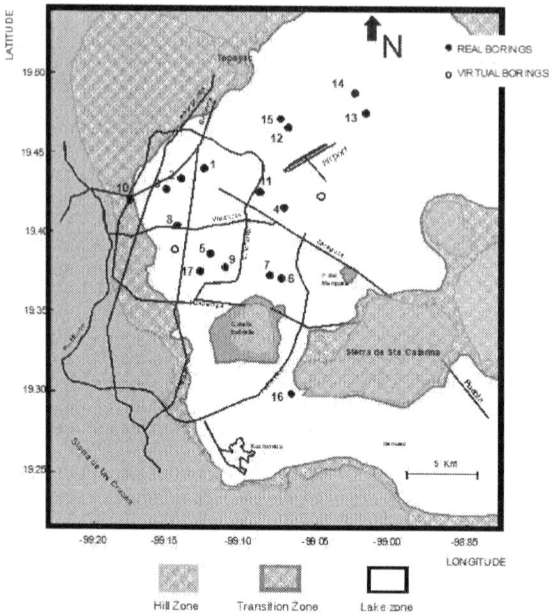

Number	SITE
1	Tlatelolco
2	Alameda
3	Plaza Córdoba
4	Velódromo
5	SCT
6	CAF
7	CDAO
8	CUPJ
9	Eugenia
10	El Águila
11	Línea B
12	Av. 510
13	Calle Urano
14	Plaza Aragón
15	Río Remedios
16	Tláhuac
17	5 de Febrero

Figure 2. Mexico City Zonation

In the 3D-neurogenetic analysis, the functions qc = {qc(X,Y,Z)}/ Vs = {Vs (X,Y,Z)} are to be approximated using the procedure outlined below:

1. Generate the database including identification of the site [borings or stations] (X,Y –geographical coordinates, Z –depth, and a CODE –ID number), elevation reference (meters above de sea level, m.a.s.l.), thickness of predetermined structures (layers), and additional information related to geotechnical zoning that could be useful for results interpretation.
2. Use the database to train an initial neural topology whose weights and layers are tuned by an evolutive algorithm (see [46] for details), until the minimum error between calculated and measured values qc=fNN(X,Y,Z)}/Vs={fNN(X,Y,Z)}is achieved (Figure 3a). The generalization capabilities of the optimal 3D neural model are tested presenting real work cases (information from borings not included in the training set) to the net. Figure 3b presents the comparison between the measured qc, Vsvalues and the NN calculations for testing cases. Through the neurogenetic results for unseen situations we can conclude that the procedure works extremely

well in identifying the general trend in materials resistance (stiffness). The correlation between NN calculations and “real” values is over 0.9.

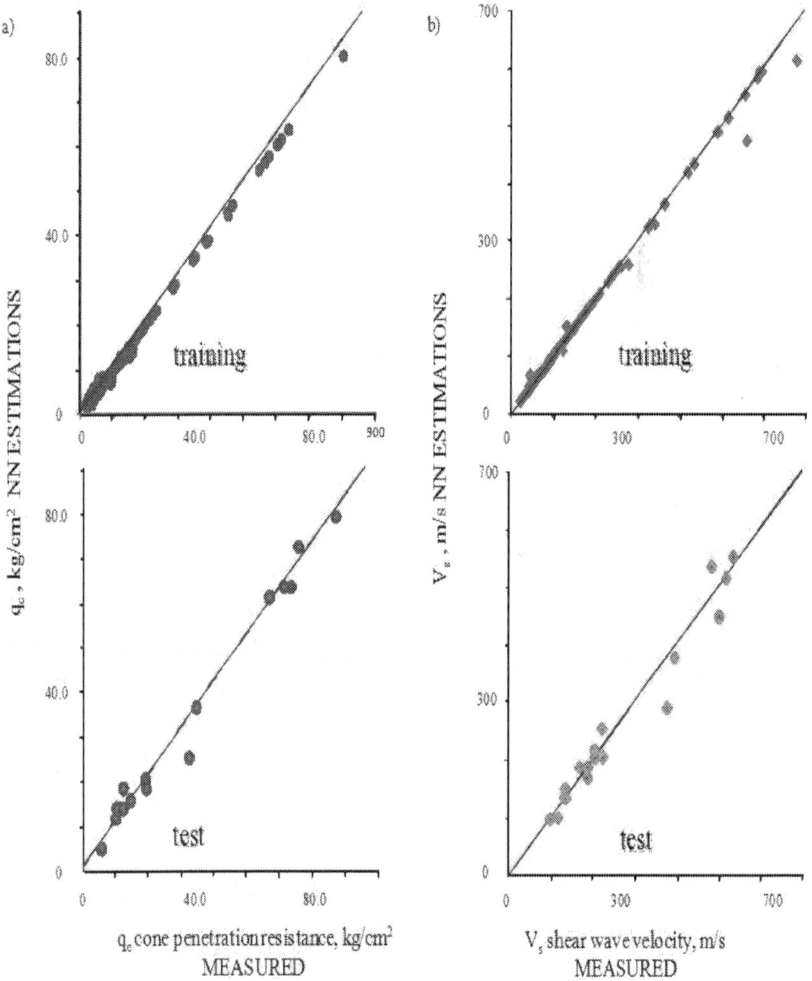

3.

Figure 3. Neural estimations of mechanical and dynamic parameters

1. For visual environment requirements a grid is constructed using raw information and neurogenetic estimations for defining the spatial variation of properties (Figure 4). The 3D view of the studied zone represents an easier and more understandable engineering system. The 3D neurogenetic-database also permits to display property-contour lines for specific depths. Using the neurogenetic contour maps, the

spatial distribution of the mechanical/dynamic variables can be visually appreciated. The 3D model is able to reflect the stratigraphical patterns (Figure 5), indicating that the proposed networks are effective in site characterization with remarkable advantages if comparing with geostatistical approximations: it is easier to use, to understand and to develop graphical user interfaces. The confidence and practical advantages of the defined neurogenetic layers is evident. Precision of predictions depends on neighborhood structure, grid size, and variance response, but based on the results we can conclude that despite of the grid cell (size) is not too small the spatial correlation extends beyond the training neighborhood, but the higher confidence is obviously only within.

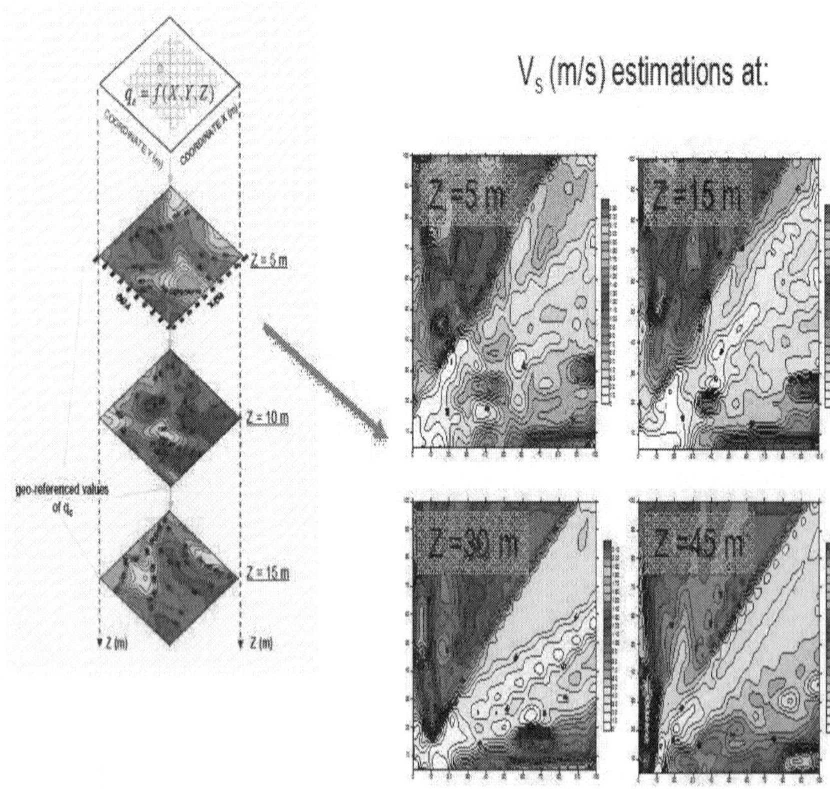

Figure 4. Neural response

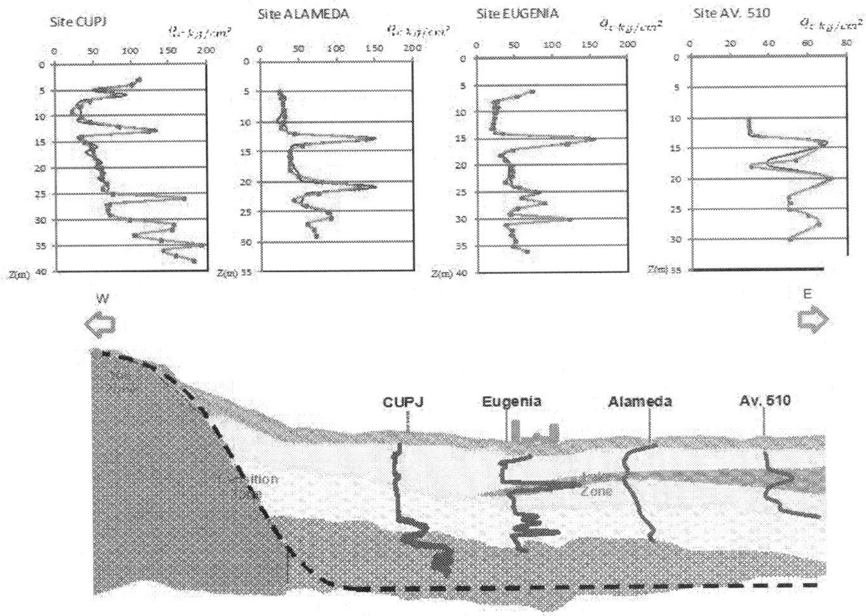

Figure 5. Stratigraphy sequence obtained using the 3D Neural estimations

Attenuation Laws for Rock Site (Outcropping Motions)

Source, path, and local site response are factors that should be considered in seismic hazard analyses when using attenuation relations. These relations, obtained from statistical regression, are derived from strong motion recordings to define the occurrence of an earthquake with a specific magnitude at a particular distance from the site. Because of the uncertainties inherent in the variables describing the source (e.g. magnitude, epicentral distance, focal depth and fault rupture dimension), the difficulty to define broad categories to classify the site (e.g. rock or soil) and our lack of understanding regarding wave propagation processes and the ray path characteristics from source to site, commonly the predictions from attenuation regression analyses are inaccurate. As an effort to recognize these aspects, multiparametric attenuation relations have been proposed by several researchers [47-53]. However, most of these authors have concluded that the governing parameters are still source, ray path, and site conditions. In this section an empirical NN formulation that uses the minimal information about magnitude, epicentral distance, and focal depth for subduction-zone earthquakes is developed to predict the peak ground acceleration PGA and spectral accelerations Sa at a rock-like site in Mexico City.

The NN model was training from existing information compiled in the Mexican strong motion database. The NN uses earthquake moment magnitude Mw, epicentral distance ED, and focal depth FD from hundreds of events recorded during Mexican subduction earthquakes (Figure 6) from 1964 to 2007. To test the predicting capabilities of the neuronal model, 186 records were excluded from the data set used in the learning phase. Epicentral distance ED is considered to be the length from the point where fault-rupture starts to the recording site, and the focal depth FD is not declared as mechanism classes, the NN should identify the event type through the FD crisp value coupled with the others input parameters [54, 47, 55], The interval of Mw goes from 3 to 8.1 approximately and the events were recorded at near (a few km) and far field stations (about 690 km). The depth of the zone of energy release ranged from very shallow to about 360 km.

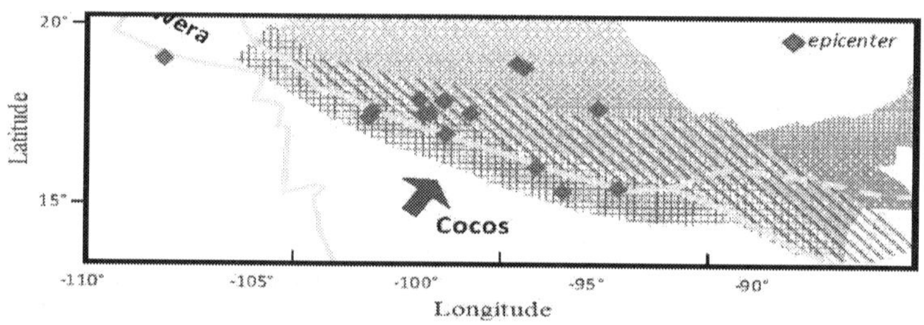

EVENT SUMMARY	
Events	80 from 1964 to 2007
Epicentral coordinates	Latitude from 13.98° to 18.74°
	Longitude from 92.79° to 104.67°
Magnitudes	from 3.9 to 8.1
Focaldepth	from <3 to 360 km
Epicentral distance	from 112 to 690 km

Figure 6. Earthquakes characteristics.

Modeling of the data base has been performed using backpropagation learning algorithm. Horizontal (mutually orthogonalPGAh1, N-S component, andPGAh2, E-W component) and vertical components (PGAv) are included as outputs for neural mapping. After trying many topologies, the best horizontal and vertical modules with quite acceptable approximations were the simpler alternatives (BP backpropagation, 2 hidden layers/15 units or nodes each). The neuronal attenuation model for {Mw,ED,FD}→{PGAh1,PGAh2,PGAv} was evaluated by performing testing analyses. The predictive capabilities of the NNs were verified by comparing the estimated PGA's to those induced by the 186 events excluded from the original database (data for training stage). In Figure 7 are compared the computed PGA's during training and testing stages to the measured values. The relative correlation factors (R2≆0.97), obtained in the training phase, indicate that those topologies selected as optimal behave consistently within the full range of intensity, distances and focal depths depicted by the patterns. Once the networks converge to the selected stop criterion, learning is finished and each of these black-boxes become a nonlinear multidimensional functional. Following this procedure 20 NN are trained to evaluate de Sa at different response spectra periods (from T= 0.1 s to T= 5.0 s with DT=0.25 s). Forecasting of the spectral components is reliable enough for practical applications.

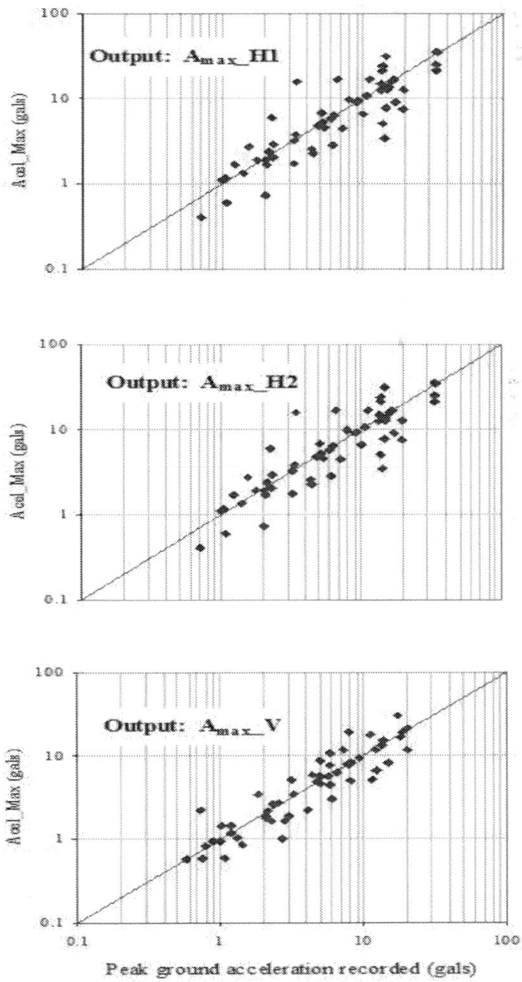

Figure 7. Some examples of measured and NN-estimated PGA values

In Figure 8 two case histories correspond to large and medium size events are shown, the estimated values obtained for these events using the relationships proposed by Gómez, Ordaz &Tena [56], Youngs et al. [47], Atkinson and Boore [55] –proposed for rock sites– and Crouse et al. [51] –proposed for stiff soil sites– and the predictions obtained with the PGAh1−h2modules are shown. It can be seen that the estimation obtained with Gómez, Ordaz y Tena [56] seems to underestimate the response for the large magnitude event. However, for the lower magnitude event follows closely both the measured responses and NN predictions. Youngs et al. [47] attenuation relationship follows closely the overall trend but tends to fall sharply for long epicentral distances.

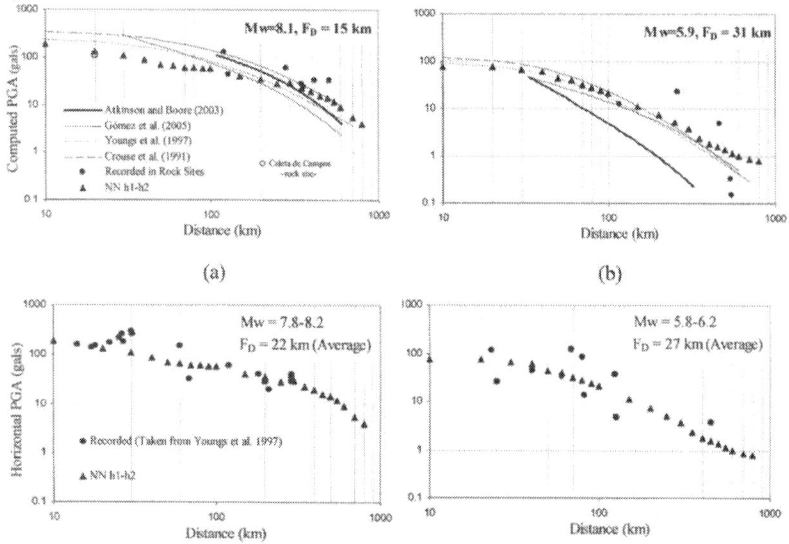

Figure 8. Attenuation laws comparisons

Furthermore, it should be stressed the fact that, as can be seen in Figure 9 the neural attenuation model is capable to follow the general behavior of the measure data expressed as spectra while the traditional functional approaches are not able to reproduce. A neural sensitivity study for the input variables was conducted for the neuronal modules. The results are strictly valid only for the data base utilized, nevertheless, after several sensitivity analyses conducted changing the database composition, it was found that the following trend prevails; theMw would be the most relevant parameter then would follow ED coupled with FD. However, for near site events the epicentral distance could become as relevant as the magnitude, particularly, for the vertical component and for minor earthquakes (M low) the FD becomes very transcendental.

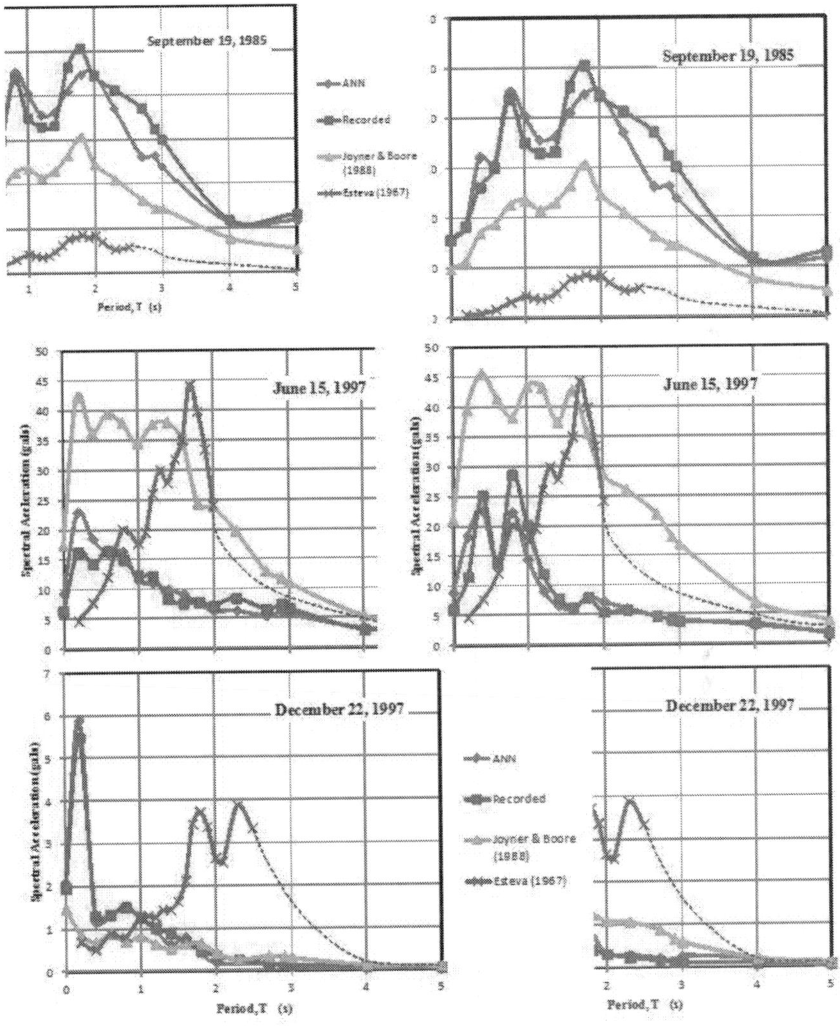

Figure 9. Response Spectra: NN-calculated vs traditional functions

Through {Mw,ED,FD}→{PGAhi,Sa} mapping, this neuronal approach offers the flexibility to fit arbitrarily complex trends in magnitude and distance dependence and to recognize and select among the tradeoffs that are present in fitting the observed parameters within the range of magnitudes and distances present in data. This approach seems to be a promising alternative to describe earthquake phenomena despite of the limited observations and qualitative knowledge of the recording stations geotechnical site conditions, which leads to a reasoning of a partially defined behavior.

Generation of Artificial Time Series: Accelerograms Application

For nonlinear seismic response analysis, where the superposition techniques do not apply, earthquake acceleration time histories are required as inputs. Virtually all seismic design codes and guidelines require scaling of selected ground motion time histories so that they match or exceed the controlling design spectrum within a period range of interest. Considerable variability in the characteristics of the recorded strong-motions under similar conditions may still require a characterization of future shaking in terms of an ensemble of accelerograms rather than in terms of just one or two "typical" records. This situation has thus created a need for the generation of synthetic (artificial) strong-motion time histories that simulate realistic ground motions from different points of views and/or with different degrees of sophistication. To provide the ground motions for analysis and design, various methods have been developed: i) frequency-domain methods where the frequency content of recorded signals is manipulated [57-60] and ii) time-domain methods where the recorded ground motions amplitude is controlled [61, 62]. Regardless of the method, first, one or more time histories are selected subjectively, and then scaling mechanisms for spectrum matching are applied. This is a trial and error procedure that leads artificial signals very far from real-earthquake time series.

In this investigation a Genetic Generator of Signals is presented. This genetic generator is a tool for finding the coefficients of a pre-specified functional form, which fit a given sampling of values of the dependent variable associated with particular given values of the independent variable(s). When the genetic generator is applied to synthetic accelerograms construction, the proposed tool is capable of i) searching, under specific soil and seismic conditions (within thousands of earthquake records) and recommending a desired subset that better match a target design spectrum, and ii) through processes that mimic mating, natural selection, and mutation, producing new generations of accelerograms until an optimum individual is obtained. The procedure is fast and reliable and results in time series that match any type of target spectrum with minimal tampering and deviation from recorded earthquakes characteristics.

The objective of the genetic generator, when applied to synthetic earthquakes construction, is to produce compatible artificial signals with specific design spectra. In this model specific seismic (fault rupture, magnitude, distance, focal depth) and site characteristics (soil/ rock) are the first set of inputs. They are included to take into consideration that a typical strong motion record consists of a variety of waves whose contribution depends on the earthquake source mechanism (wave path) and its particular characteristics are influenced by the distance between the source and the site, some measure of the size of the earthquake, and the surrounding geology and site conditions; and that the design spectra can be an envelope or integration of many expected ground motions that are possible to occur in certain period of time, or the result of a formulation that involves earthquake magnitude, distance and soil conditions. The second set of inputs consist of the target spectrum, the period range for the matching, lower- and upper-bound acceptable values for scaling signal shape, and a collection of GAs parameters (a population size, number of generations, crossover ratio, and mutation ratio). The output is the more success individual with a chromosome array generated from "real" accelerograms parents (a set of).

The algorithm (see Figure 10) is started with a set of solutions (each solution is called a chromosome). A solution is composed of thousands of components or genes (accelerations recorded at the time), each one encoding a particular trait. The initial solutions (original population) are selected based on the seismic parameters at a site (defined previously by the user): fault mechanism, moment magnitude, epicentral distance, focal depth, geotechnical and geological site classification, depth of sediments. If the user does not have a priori seismic/site knowledge, the genetic generator could select the initial population randomly (Figure 11). Once the model has found the seed-accelerogram(s) or chromosome(s), the space of all feasible solutions can be called accelerograms space (state space). Each point in this search space represents one feasible solution and can be "marked" by its value or fitness for the problem. The looking for a solution is then equal to a looking for some extreme (minimum or maximum) in the space.

According to the individuals' fitness, expressed by difference between the target design spectrum and the chromosome response spectrum, the problem is formulated as the minimization of the error function, Z, between the actual and the target spectrum in a certain period range. Solutions with highest fitness are selected to form new solutions (offspring). During reproduction, the recombination (or crossover) and mutation permits to change the genes (accelerations) from parents (earthquake signals) in some way that the whole new chromosome (synthetic signal) contains the older organisms attributes that assure success. This is repeated until some user's condition (for example number of populations or improvement of the best solution) is satisfied (Figure 12).

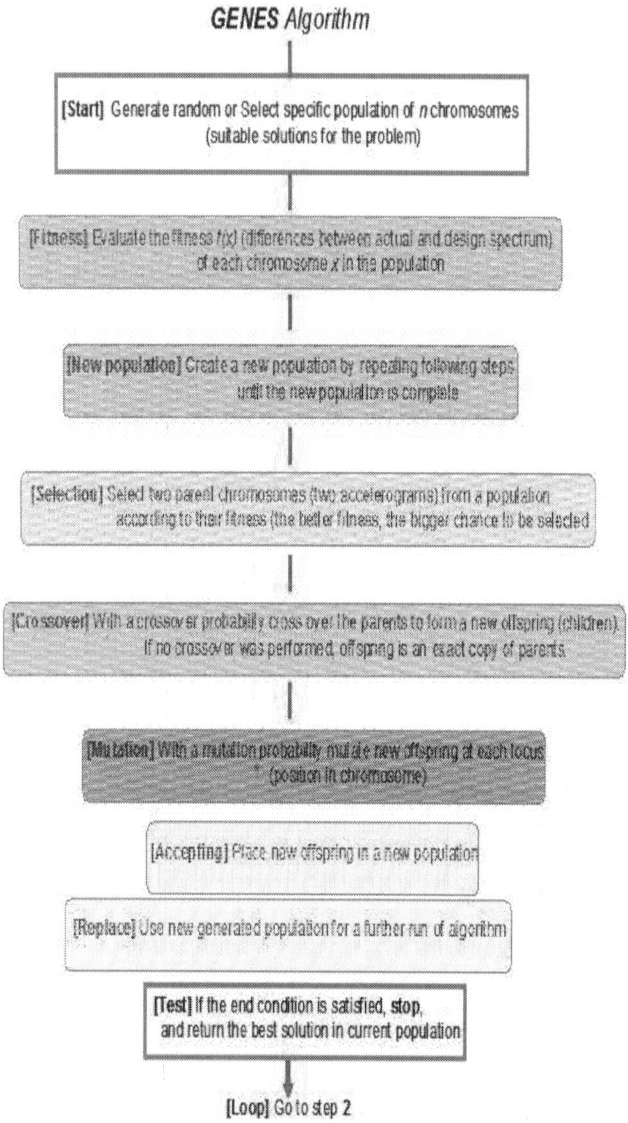

Figure 10. Genetic Generator: flow diagram

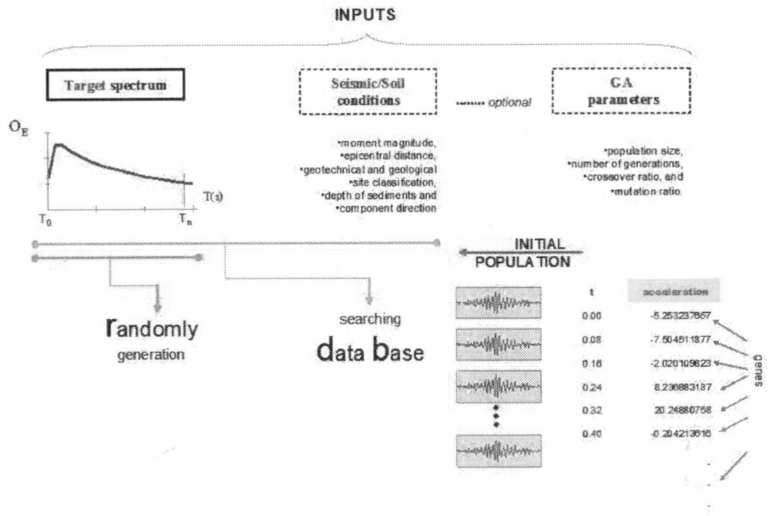

Figure 11. Genetic Generator: working phase diagram

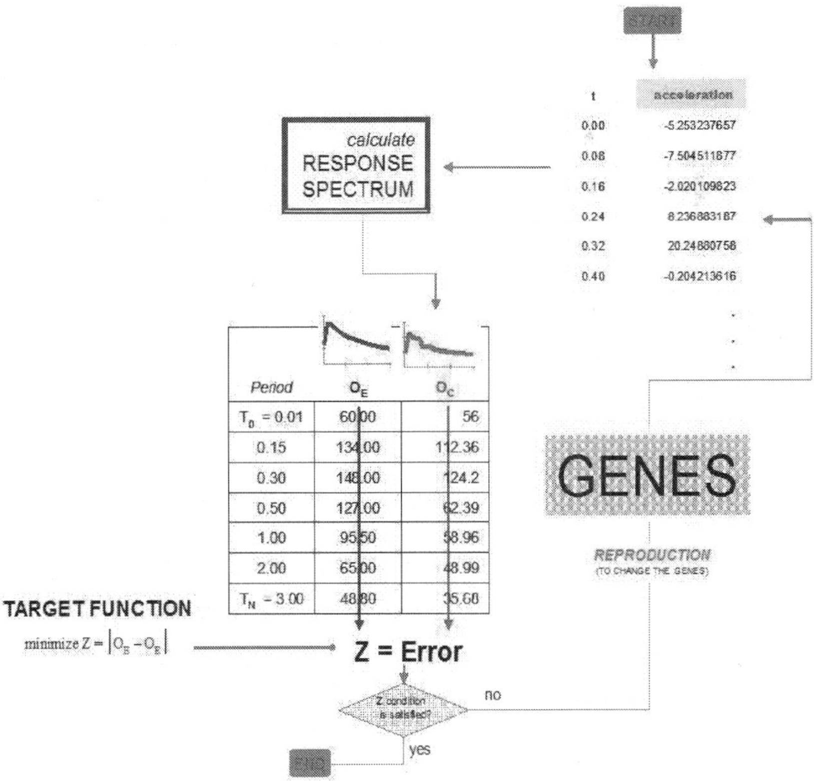

Figure 12. Iteration process of the Genetic Generator

One of the genetic advantages is the possibility of modifying on line the image of the expected earthquake. While the genetic model is running the user interface shows the chromosome per epoch and its response spectra in the same window, if the duration time, the highest intensities interval or the Øtare not convenient for the user's interests, these values can be modified without retraining or a change on model' structure.

In Figure 13 are shown three examples of signals recovered following this methodology. The examples illustrate the application of the genetic methodology to select any number of records to match a given target spectrum (only the more successful individuals for each target are shown in the figure). It can be noticed the stability of the genetic algorithm in adapting itself to smooth, code or scarped spectrum shapes. The procedure is fast and reliable as results in records match the target spectrum with minimal deviation. The genetic procedure has been applied successfully to generate synthetic ground motions having different amplitudes, duration and combinations of moment magnitude and epicentral distance. Although the variations in the target spectra, the genetic signals maintain the nonlinear and nonstationary characteristics of real earthquakes. It is still under development an additional toolbox that will permit to use advanced signal analysis instruments because, as it has been demonstrated [63] [64], studying nonstationary signals through Fourier or response spectra is not convenient for all applications.

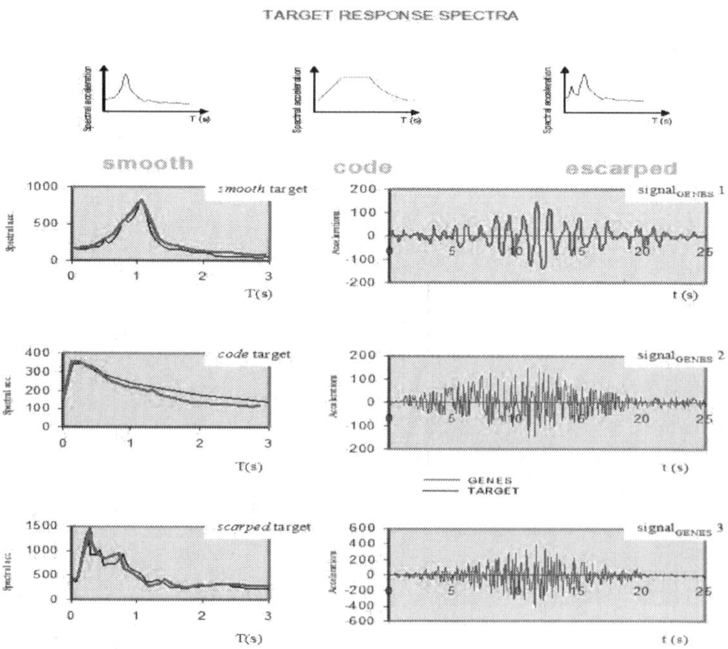

Figure 13. Some Generator results: accelerograms application

Effects of Local Site Conditions on Ground Motions

Geotechnical and structural engineers must take into account two fundamental characteristics of earthquake shaking: 1) how ground shaking propagates through the Earth, especially near the surface (site effects), and 2) how buildings respond to this ground motion. Because neither characteristic is completely understood, the seismic phenomenon is still a challenging research area.

Site effects play a very important role in forecasting seismic ground responses because they may strongly amplify (or deamplify) seismic motions at the last moment just before reaching the surface of the ground or the basement of man-made structures. For much of the history of seismological research, site effects have received much less attention than they should, with the exception of Japan, where they have been well recognized through pioneering work by Sezawa and Ishimoto as early as the 1930's [65]. The situation was drastically changed by the catastrophic disaster in Mexico City during the Michoacan, Mexico earthquake of 1985, in which strong amplification due to extremely soft clay layers caused many high-rise buildings to collapse despite their long distance from the source. The cause of the astounding intensity and long duration of shaking during this earthquake is not well resolved yet even though considerable research has been conducted since then, however, there is no room for doubt that the primary cause of the large amplitude of strong motions in the soft soil (lakebed) zone relative to those in the hill zone is a site effect of these soft layers.

The traditional data-analysis methods to study site effects are all based on linear and stationary assumptions. Unfortunately, in most soil systems, natural or manmade ones, the data are most likely to be both nonlinear and nonstationary. Discrepancies between calculated responses (using code site amplification factors) and recent strong motion evidence point out serious inaccuracies may be committed when analyzing amplification phenomena. The problem might be due partly because of the lack of understanding regarding the fundamental causes in soil response but also a consequence of the distorted soil amplification quantification and the incomplete characterization of nonlinearity-induced nonstationary features exposed in motion recordings [66]. The objective of this investigation is to illustrate a manner in which site effects can be dealt with for the case of Mexico City soils, making use of response spectra calculated from the motions recorded at different sites during extreme and minor events (see Figure 6). The variations in the spectral shapes, related to local site conditions, are used to feed a multilayer neural network that represent a very advantageous nonlinear-amplification relation. The database is composed by registered information earthquakes affecting Mexico City originated by different source mechanisms.

The most damaging shocks, however, are associated to the subduction of the Cocos Plate into the Continental Plate, off the Mexican Pacific Coast. Even though epicentral distances are rather large, these earthquakes have recurrently damaged structures and produced severe losses in Mexico City. The singular geotechnical environment that prevails in Mexico City is the one most important factor to be accounted for in explaining the huge amplification of seismic movements [67-70]. The soils in Mexico City were formed by the deposition into a lacustrine basin of air and water transported materials. From the view point of geotechnical engineering, the relevant strata extend down to depths of 50 m to 80 m, approximately. The superficial layers formed the bed of a lake system that has been subjected to dessication for the last 350 years. Three types of soils may be broadly distinguished: in Zone I, firm soils and rock-like materials prevail; in Zone III, very soft clay formations with large amounts of microorganisms interbedded by thin seams of silty sands, fly ash and volcanic glass are found; and in Zone II, which is a transition between zones I and III, sequences of clay layers an coarse material strata are present (Figure 14).

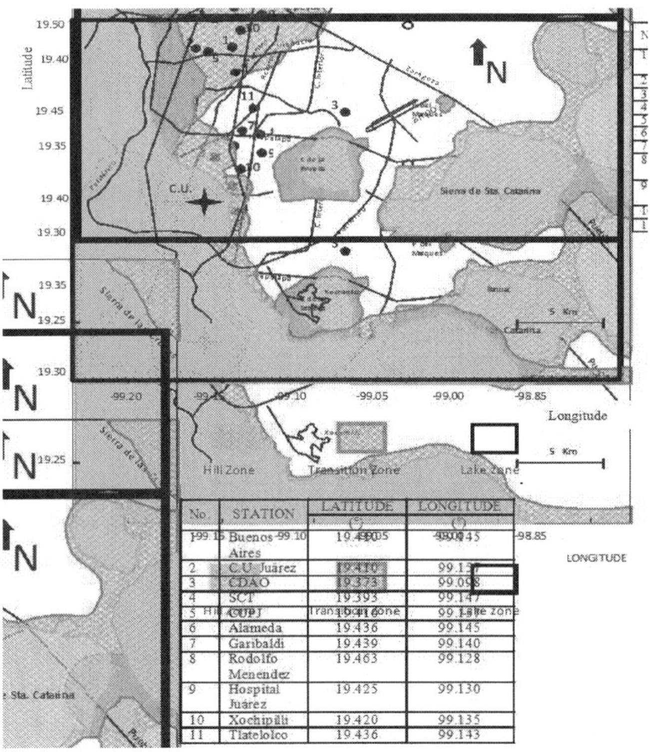

No.	STATION	LATITUDE (°)	LONGITUDE (°)
1	Buenos Aires	[illegible]	[illegible]
2	C.U. Juárez	19.410	[illegible]
3	CDAO	19.373	[illegible]
4	SCT	19.393	99.147
5	[illegible]	[illegible]	[illegible]
6	Alameda	19.436	99.145
7	Garibaldi	19.439	99.140
8	Rodolfo Menendez	19.463	99.128
9	Hospital Juárez	19.425	99.130
10	Xochipilli	19.420	99.135
11	Tlatelolco	19.436	99.143

Figure 14. Accelerographic stations used in this study

Due to space limitations, reference is made only to two seismic events: the June 15, 1999 and the October 9, 1995. This module was developed based in a previous study (see section 4.2 of this Chapter) where the effect of the parameters ED, FD and Mw on the ground motion attenuation from epicentre to the site, were found to be the most significant [71]. The recent disaster experience showed that the imprecision that is inherent to most variables measurements or estimations makes crucial the consideration of subjectivity to evaluate and to derive numerical conclusions according to the phenomena behavior. The neuronal training process starts with the training of four input variables booked: ED, FD and Mw. The output linguistic variables are PGAh1 (peak ground acceleration horizontal, component 1) and ,PGAh2 (peak ground acceleration, horizontal component 2) registered in a rock-like site in Zone I .The second training process is linked *feed-forward* with the previous module (PGA for rock-like site) and the new seismic inputs are Seismogenic Zone and PGA_{rock} and the Latitude and Longitude coordinates are the geo-referenced position needed to draw the deposition variation into the basin. This neuro-training runs one step after the first training phase and until the minimum difference between theSa and the neuronal calculations is attained. In Figure 15 some results from training and testing modes are shown.

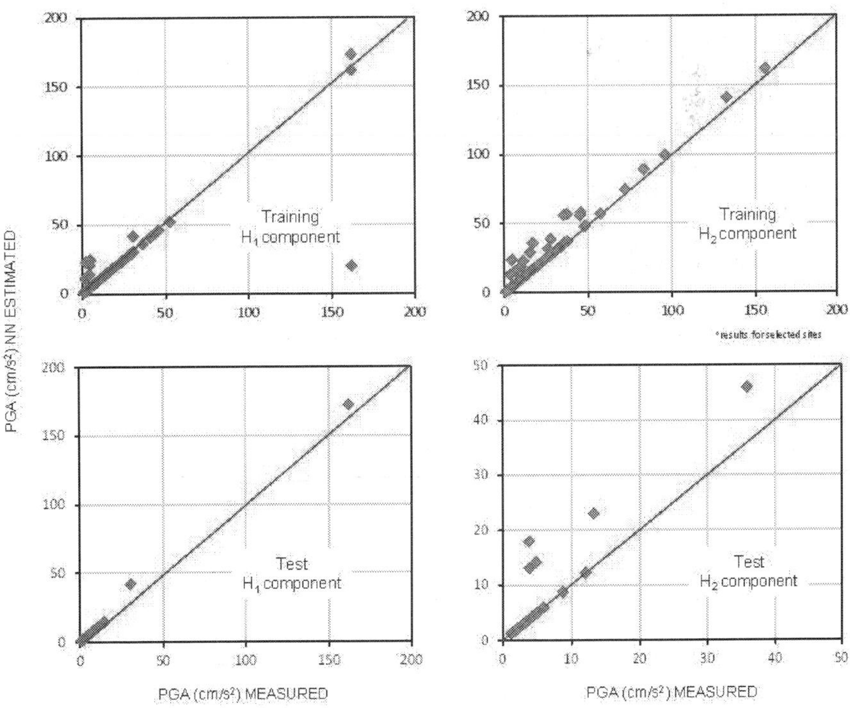

Figure 15. Neural estimations for PGA in Lake Zone site

This second NN represents the geo-referenced amplification ratio that take into consideration the topographical, geotechnical and geographical conditions, implicit in the recorded accelerograms.The results of these two NNs are summarized in Figure 16. These graphs show the predicting capabilities of the neural system comparing the measured values with those obtained in neural-working phase. It can be observed a good correspondence throughout the full distance and magnitude range for the seismogenic zones considered in this study for the whole studied area (Lake Zone).

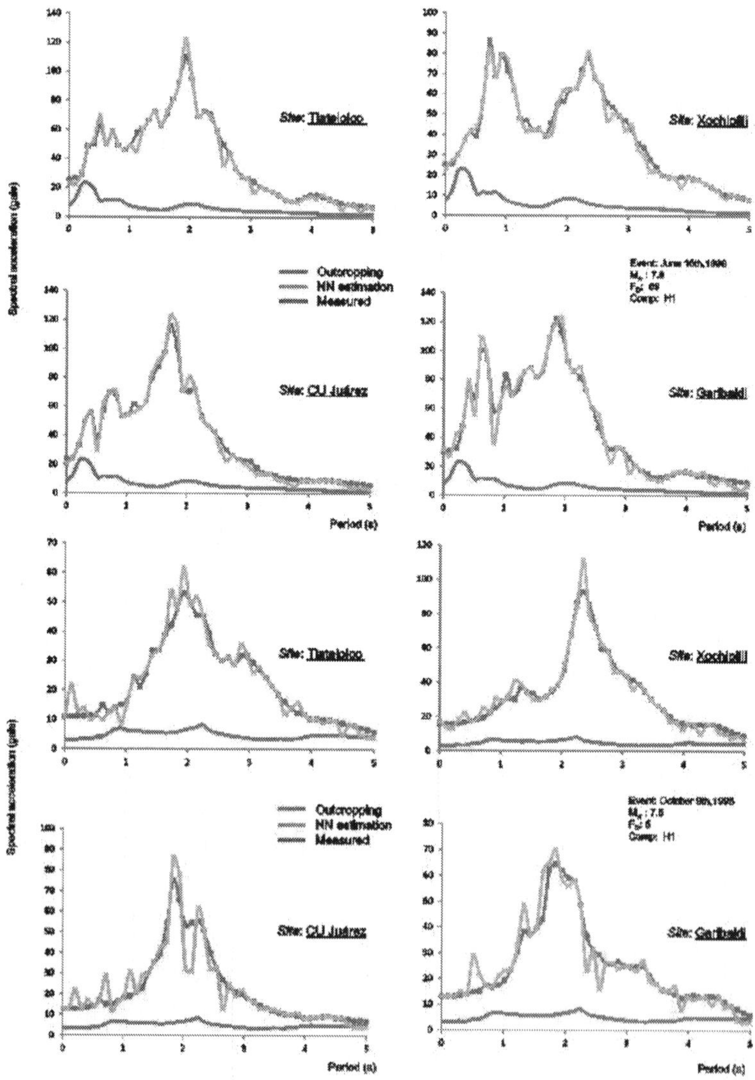

Figure 16. Spectral accelerations in some Lake-Zone sites: measured vs NN

Liquefaction Phenomena: Potential Assessment and Lateral Displacements Estimation

Over the past forty years, scientists have conducted extensive research and have proposed many methods to predict the occurrence of liquefaction. In the beginning, undrained cyclic loading laboratory tests had been used to evaluate the liquefaction potential of a soil [72] but due to difficulties in obtaining undisturbed samples of loose sandy soils, many researchers have preferred to use *in situ* tests [73]. In a semi-empirical approach the theoretical considerations and experimental findings provides the ability to make sense out of the field observations, tying them together, and thereby having more confidence in the validity of the approach as it is used to interpolate or extrapolate to areas with insufficient field data to constrain a purely empirical solution. Empirical field-based procedures for determining liquefaction potential have two critical constituents: i) the analytical framework to organize past experiences, and ii) an appropriate *in situ* index to represent soil liquefaction characteristics. The original simplified procedure [74] for estimating earthquake-induced cyclic shear stresses continues to be an essential component of the analysis framework. The refinements to the various elements of this context include improvements in the in-situ index tests (e.g., SPT, CPT, BPT,Vs), and the compilation of liquefaction/no-liquefaction cases.

The objective of the present study is to produce an empirical machine learning ML method for evaluating liquefaction potential. ML is a scientific discipline concerned with the design and development of algorithms that allow computers to evolve behaviours based on empirical data, such as from sensor data or databases. Data can be seen as examples that illustrate relations between observed variables. A major focus of ML research is to automatically learn to recognize complex patterns and make intelligent decisions based on data; the difficulty lies in the fact that the set of all possible behaviours given all possible inputs is too large to be covered by the set of observed examples (training data). Hence the learner must generalize from the given examples, so as to be able to produce a useful output in new cases. In the following two ML tools, Neural Networks NN and Classification Trees CTs, are used to evaluate liquefaction potential and to find out the liquefaction control parameters, including earthquake and soil conditions. For each of these parameters, the emphasis has been on developing relations that capture the essential physics while being as simplified as possible. The proposed cognitive environment permits an improved definition of i) seismic loading or cyclic stress ratio CSR, and ii) the *resistance* of the soil to triggering of liquefaction or cyclic resistance ratio CRR.

The factor of safety FS against the initiation of liquefaction of a soil under a given seismic loading is commonly described as the ratio of cyclic resistance ratio (CRR), which is a measure of liquefaction resistance, over cyclic stress ratio (CSR), which is a representation of seismic loading that causes liquefaction, symbolically, FS=CRR/CSR. The reader is referred to Seed and Idriss [74], Youd et al.[75], and Idriss and Boulanger [76] for a historical perspective of this approach. The term CSR CSR = f (0.65,σ vo,amax,σ' vo, rd, MSF)is function of the vertical total stress of the soil σvo at the depth considered, the vertical effective stress σ' vo, the peak horizontal ground surface acceleration amax, a depth-dependent shear stress reduction factor rd (dimensionless), a magnitude scaling factor MSF(dimensionless). For CRR, different in situ-resistance measurements and overburden correction factors are included in its determination; both terms operate depending of the geotechnical conditions. Details about the theory behind this topic in Idriss and Boulanger, [76] and Youd et al. [75].

Many correction/adjustment factors have been included in the conventional analytical frameworks to organize and to interpret the historical data. The correction factors improve the consistency between the geotechnical/seismological parameters and the observed liquefaction behavior, but they are a consequence of a constrained analysis space: a 2D plot [CSR *vs.* CRR] where regression formulas (simple equations) intend to relate complicated nonlinear/multidimensional information. In this investigation the ML methods are applied to discover unknown, valid patterns and relationships between geotechnical, seismological and engineering descriptions using the relevant available information of liquefaction phenomena (expressed as empirical prior knowledge and/or input-output data). These ML techniques "work" and "produce" accurate predictions based on few logical conditions and they are not restricted for the mathematical/analytical environment. The ML techniques establish a*natural* connection between experimental and theoretical findings.

Following the format of the simplified method pioneered by Seed and Idriss [74], in this investigation a nonlinear and adaptative *limit state* (a fuzzy-boundary that separates liquefied cases from nonliquefied cases) is proposed (Figure 17). The database used in the present study was constructed using the information included in Table 2 and it was compiled by Agdha et al., [77], Juang et al., [78], Juang [79], Baziar, [80] and Chern and Lee [81]. The cases are derived from cone penetration tests CPT, and shear wave velocities Vs measurements and different world seismic conditions (U.S., China, Taiwan, Romania, Canada and Japan). The soils types ranges from clean sand and silty sand to silt mixtures (sandy and clayey silt). Diverse geological and geomorphological characteristics are included. The reader is referred to the citations in Table 2 for details.

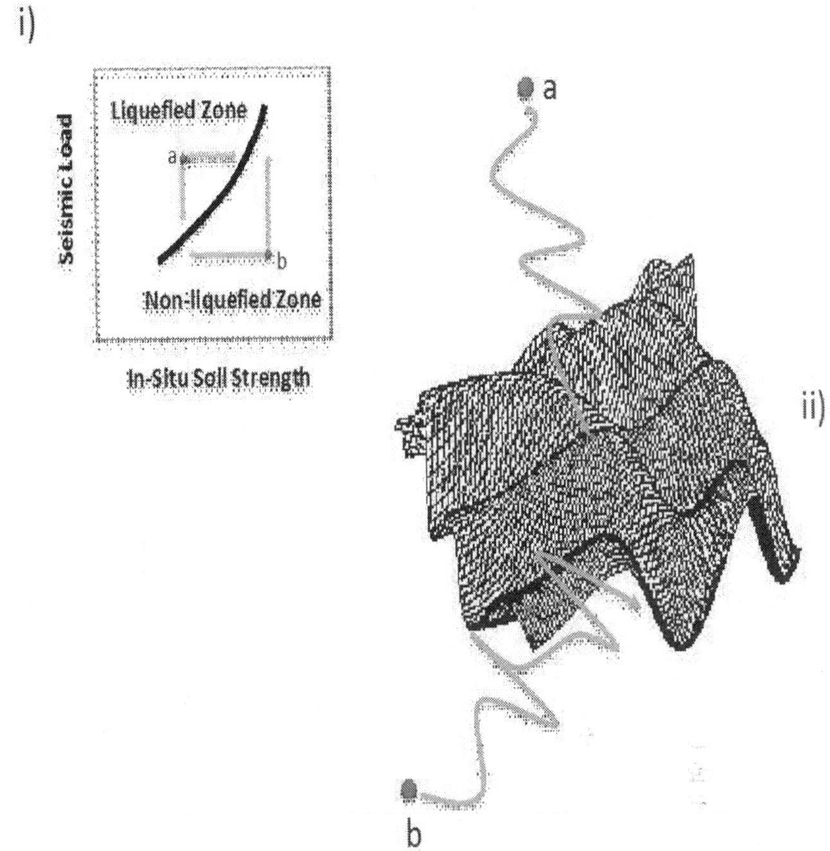

Figure 17. An schematic view of the nonlinear-liquefaction boundary

The ML formulation uses Geotechnical (qc, Vs, Unit weight, Soil Type, Total vertical stresses, Effective vertical stresses, Geometrical (Layer thickness, Water Level Depth, Top Layer Depth) andSeismological (Magnitude, PGA) input parameters and the output variable is "Liquefaction?" and it can take the values "YES/NO" (Figure 17). Once the NN is trained the number of cases that was correctly evaluated was 100% and applied to "unseen" cases (separated for testing) less than 10% of these examples were not fitted. The CT has a minor efficiency during the training showing 85% of cases correctly predicted, but when the CT runs on the unseen patterns its capability is not diminished and it asserts the same proportion. From these findings it is concluded that the neuro system is capable of predicting the in situ measurements with a high degree of accuracy but if improvement of knowledge is necessary or there are missed, vague even contradictory values in the analyzed case, the CT is a better option.

Table 2. Database for liquefaction analysis

Set	Input Parameters	Number .of Patterns	Ref.
A	Z, Z_{NAF}; H, Soil Class, Geomorphological units, Geological units, Site amplification, a_{max}	56	Fatemi-Agdha et al., 1988
B	Z, q_c, F_s, σ_0, σ_0', a_{max}, M	21	Juang et al., 1999
C	Z, q_c, F_s, σ_0, σ_0', a_{max}, M	242	Juang, 2003
D	D_{50}, a_{max}, σ_0', σ_0, M, F_s, q_c, SPT, Z	170	Baziar, 2003
E	M, σ_0, σ_0', q_c, a_{max}	466	Chern and Lee, 2009
F	Z_{NAF}, Z, H, σ_0, σ_0', Soil Class, V_s	80	Andrus and Stokoe, 1997; 2000
	Total:	**1035**	

Figure 18 shows the pruned liquefaction trees (two, one runs using qc values and the other through theVs measurements) with YES/NO as terminal nodes. In the Figure 19, some examples of tree reading are presented. The trees incorporate soil type dependence through the resistance values (qc, andVs) and fine content, and it is not necessary to label the material as "sand" or "silt". The most general geometrical branches that split the behaviors are the Water table depth and the Layer thickness but only when the soil description is based onVs, whenqc, serves as rigidity parameter this geometrical inputs are not explicit exploited. This finding can be related to the nature of the measurement: the cone penetration value contains the effect of the saturated material while the shear wave velocities need the inclusion of this situation explicitly. Without potentially confusing regression strategies, the liquefaction trees results can be seen as an indication of how effectively the ML model maps the assigned predictor variables to the response parameter. Using data from all regions and wide parameters ranges, the prediction capabilities of the neural network and classification trees are superior to many other approximations used in common practice, but the most important remark is the generation of meaningful clues about the reliability of physical parameters, measurement and calculation process and practice recommendations.

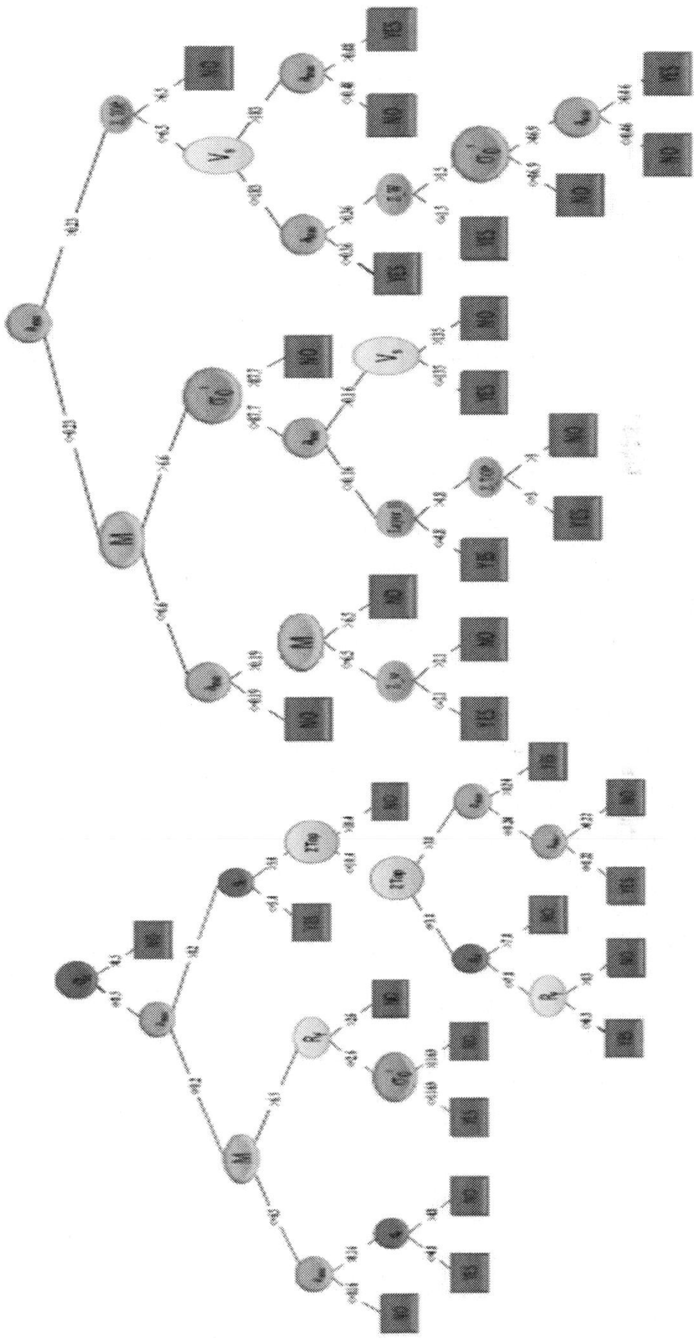

Figure 18. Classification tree for liquefaction potential assessment

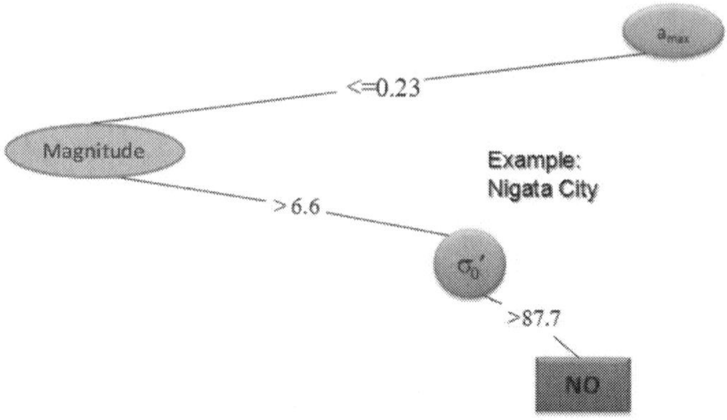

Magnitude (Mw)	Depth (m)	Z_w (m)	Z_TOP (m)	Layer_H (m)	σ'v (kPa)	R_f (%)	Type of Soil	qc (MPA)	Vs (m/s)	amax (g)	Liquefied
7.50	-	5.00	5.00	2.50	97.70	-	silt	-	163.00	0.16	no

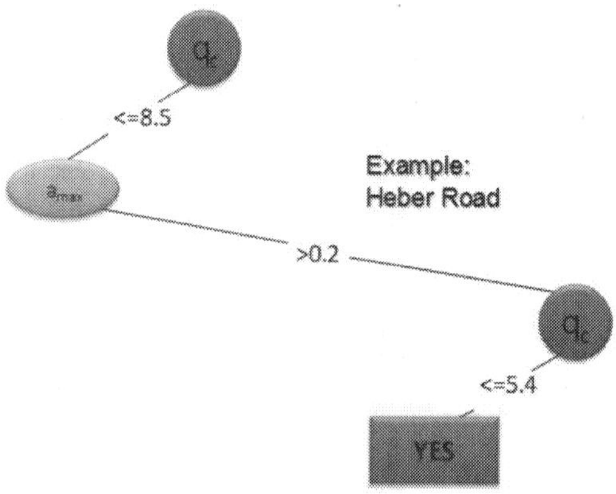

Magnitude (Mw)	Depth (m)	Z_w (m)	Z_TOP (m)	Layer_H (m)	σ'v (kPa)	R_f (%)	Type of Soil	qc (MPA)	Vs (m/s)	amax (g)	Liquefied
6.60	4.00	-	-	-	56.00	2.80	-	2.0	-	0.8	yes

Figure 19. CT classification examples

The intricacy and nonlinearity of the phenomena, an inconsistent and contradictory database, and many subjective interpretations about the observed behavior, make SC an attractive alternative for estimation of liquefaction induced lateral spread. NEFLAS [82], NEuro Fuzzy estimation of liquefaction induced LAteral Spread, profits from fuzzy and neural paradigms through an architecture that uses a fuzzy system to represent knowledge in an interpretable manner and proceeds from the learning ability of a neural network to optimize its parameters. This blending can constitute an interpretable model that is capable of learning the problem-specific prior knowledge.

NEFLAS is based on the Takagi-Sugeno model structure and it was constructed according the information compiled by Bartlett and Youd [83] and extended later by Youd et al. [84]. The output considered in NEFLAS is the horizontal displacements due to liquefaction, dependent of moment magnitude, the PGA, the nearest distance from the source in kilometers; the free face ratio, the gradient of the surface topography or the slope of the liquefied layer base, the cumulative thickness of saturated cohesionless sediments with number of blows (modified by overburden and energy delivered to the standard penetration probe, in this case 60%) , the average of fines content, and the mean grain size.

One of the most important NEFLAS advantages is its capability of dealing with the imprecision, inherent in geoseismic engineering, to evaluate concepts and derive conclusions. It is well known that engineers use words to classify qualities ("strong earthquake", "poor graduated soil" or "soft clay" for example), to predict and to validate "first principle" theories, to enumerate phenomena, to suggest new hypothesis and to point the limits of knowledge. NEFLAS mimics this method. See the technical quantity "magnitude" (earthquake input) depicted in Figure 20. The degree to which a crisp magnitude belongs to LOW, MEDIUM or HIGH linguistic label is called the degree of membership. Based on the figure, the expression, "the magnitude is LOW" would be true to the degree of 0.5 for aMwof 5.7. Here, the degree of membership in a set becomes the degree of truth of a statement.

On the other hand, the human logic in engineering solutions generates sets of behavior rules defined for particular cases (parametric conditions) and supported on numerical analysis. In the neurofuzzy methods the human concepts are re-defined through a flexible computational process (training) putting (empirical or analytical) knowledge into simple "if-then" relations (Figure 20). The fuzzy system uses 1) variables composing the antecedents (premises) of implications; 2) membership functions of the fuzzy sets in the premises, and 3) parameters in consequences for finding simpler solutions with less design time.

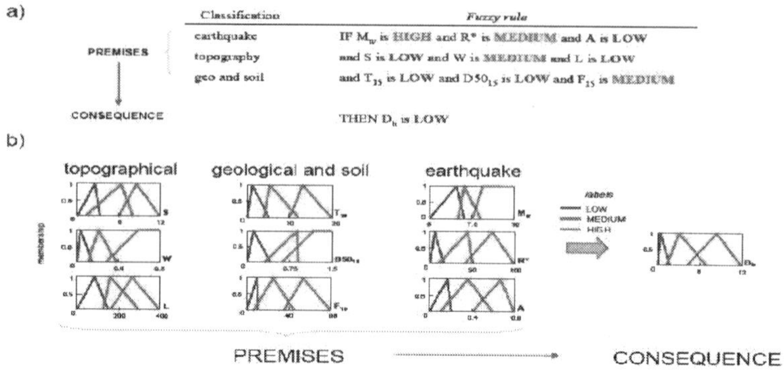

Figure 20. Neurofuzzy estimation of lateral spread

NEFLAS considers the character of the earthquake, topographical, regional and geological components that influence lateral spreading and works through three modules: Reg-NEFLAS, appropriate for predicting horizontal displacements in geographic regions where seismic hazard surveys have been identified; Site- NEFLAS, proper for predictions of horizontal displacements for site-specific studies with minimal data on geotechnical conditions and Geotech-NEFLAS allows more refined predictions of horizontal displacements when additional data is available from geotechnical soil borings. The NEFLAS execution on cases not included in the database (Figure 21.b and Figure 21.c) and its higher values of correlation when they are compared with evaluations obtained from empirical procedures permit to assert that NEFLAS is a powerful tool, capable of predicting lateral spreads with high degree of confidence.

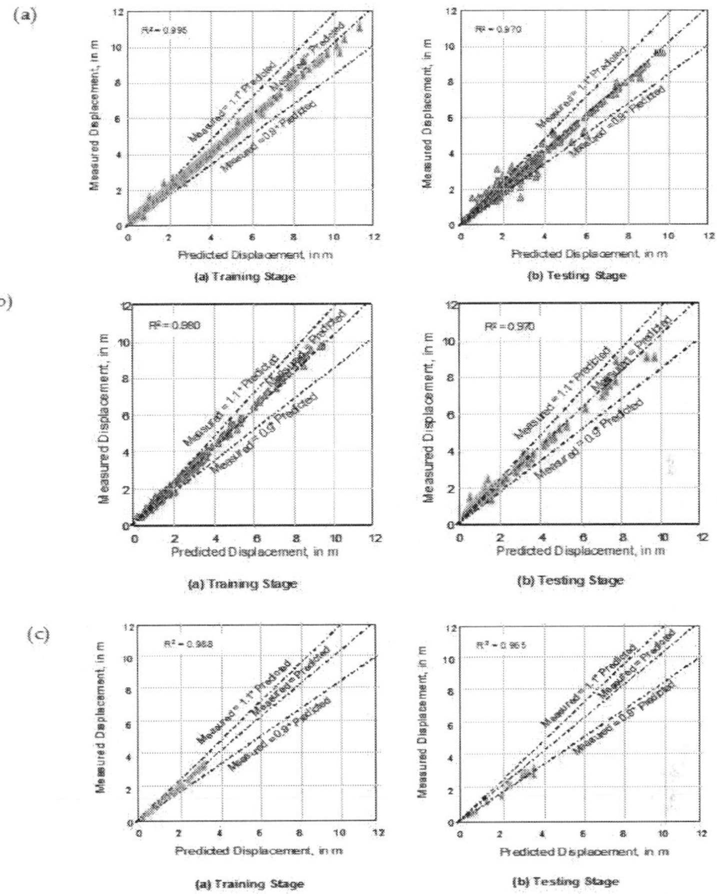

Figure 21. NN estimations vs measured displacements for a) the whole data set, b)Niigata Japan, c) San Francisco USA cases

CONCLUSIONS

Based on the results of the studies discussed in this paper, it is evident that cognitive techniques perform better than, or as well as, the conventional methods used for modeling complex and not well understood geotechnical earthquake problems. Cognitive tools are having an impact on many geotechnical and seismological operations, from predictive modeling to diagnosis and control.

The hybrid *soft* systems leverage the tolerance for imprecision, uncertainty, and incompleteness, which is intrinsic to the problems to be solved, and generate tractable, low-cost, robust solutions to such problems. The synergy derived from these hybrid systems stems from the relative ease with which we can translate problem domain knowledge into initial model structures whose parameters are further tuned by local or global search methods. This is a form of methods that do not try to solve the same problem in parallel but they do it in a mutually complementary fashion. The push for low-cost solutions combined with the need for intelligent tools will result in the deployment of hybrid systems that efficiently integrate reasoning and search techniques.

Traditional earthquake geotechnical modeling, as physically-based (or knowledge-driven) can be improved using soft technologies because the underlying systems will be explained also based on data (CC data-driven models). Through the applications depicted here it is sustained that cognitive tools are able to make abstractions and generalizations of the process and can play a complementary role to physically-based models.

REFERENCES

1. W.D.L.Finn State of the art of geotechnical earthquake engineering practice Soil Dynamics and Earthquake Engineering 20 (2000Elsevier
2. N. A. R. R. Abrahamson, A. Youngs, algorithm. stable, regression. for, using. analysis, random. the, model. effects, Bull Seismol Soc Am 1992
3. P. G. Somerville, T. Sato, Correlation of rise time with the style-offaulting factor in strong ground motions. Seismol Res Lett 1998pp (abstract).
4. Somerville PG, Greaves RL.Strong ground motions of the Kobe, Japan earthquake of January 17, 1995and development of a model of forward rupture directivity applicable in California. Proc. Western Regional Tech. Seminar on Earthquake Eng. for Dams. Assoc. of State Dam Safety Oficials, Sacramento, CA, April 11±12, 1996.
5. Abrahamson NA, Silva WJ.Empirical duration relations for shallow crustal earthquakes. Written communication, 1997
6. R. W. Graves, A. Pitarka, P. G. Somerville, Ground motion amplication in the Santa Monica area: effects of shallow basin edge structure. Submitted for publication.
7. P. G. Somerville, art. Emerging, ground. earthquake, In. motion, P. Dakoulas, M. Yegian, R. D. Holtz, editors, Proc Geotechnical Earthquake Engineering in Soil Dynamics III, Geotechnical Special Publication 751ASCE, 19981
8. Abrahamson NA.Spatial variation of multiple support inputs. Proc. First US Symp. Seism. Eval. Retrofit Steel Bridges, UC, Berkeley, October 18, 1993
9. F. Naeim, M. Lew, On the use of design spectrum compatible motions. Earthquake Spec 1995II(1):111±28.

10. Boulanger R.W. and Idriss, I.M.2006Liquefaction Susceptibility Criteria for Silts and Clays," Journal of Geotechnical and Geoenvironmental Engineering, 132 (11), 1413 EOF1426 EOFpp.
11. Boulanger R.W. and Idriss, I.M.2007Evaluation of Cyclic Softening in Silts and Clays", Journal of Geotechnical and Geoenvironmental Engineering, 133 (6), 641 EOF652 EOFpp.
12. Youd and Idriss, NCEER. Proceedings, Workshop on Evaluation of Liquefaction Resistance of Soils.Technical Report NCCER-97970022National Center for Earthquake Engineering Research, University of Buffalo, Buffalo, New York, 1997
13. H. B. Seed, K. Tokimatsu, L. F. Harder, R. M. Chung, of. S. P. T. Influence, in. procedures, liquefaction. soil, evaluations. resistance, J Geotech Engng 19851425 EOF
14. Robertson PK, Fear CE.Liquefaction of sands and its evaluation. Proceedings, 1st Int. Conf. on Earthquake Geotechnical Engineering, Tokyo, Japan, 1995
15. Ambraseys NN. Engineering seismology.Earthquake Engng Struct Dynam 1988
16. Y. Wang, 2008aOn Contemporary Denotational Mathematics for Computational Intelligence. In Transactions of Computational Science (2629New York: Springer.
17. Y. Wang, 2009aOn Abstract Intelligence: Toward a Unified Theory of Natural, Artificial, Machinable, and Computational Intelligence. International Journal of Software Science and Computational Intelligence, 1 EOF17 EOF
18. Y. Wang, 2009bOn Cognitive Computing. International Journal of Software Science and Computational Intelligence, 1(3), 1-15.
19. A. M. Turing, 1950Computing Machinery and Intelligence. Mind, 59433460
20. Neumann. J. Von, 1946The Principles of Large-Scale Computing Machines. Reprinted in Annals of History of Computers, 3(3), 263-273.
21. Neumann. J. Von, 1958The Computer and the Brain. New Haven: Yale Univ. Press.
22. J. L. Gersting, 1982Mathematical Structures for Computer Science. San Francisco: W. H. Freeman & Co.
23. D. Mandrioli, C. Ghezzi, 1987Theoretical Foundations of Computer Science. New York: John Wiley & Sons.
24. H. R. Lewis, C. H. Papadimitriou, 1998Elements of the Theory of Computation, 2nd ed. Englewood Cliffs, NJ: Prentice Hall International.
25. J. Kephart, D. Chess, 2003The Vision of Autonomic Computing. IEEE Computer, 26(1), 41 EOF50 EOF
26. IBM2006Autonomic Computing White Paper. An Architectural Blueprint for Autonomic Computing, 4th ed., June, (137

27. Y. Wang, 2004Keynote: On Autonomic Computing and Cognitive Processes. Proc. 3rd IEEE International Conference on Cognitive Informatics (ICCI'04), Victoria, Canada, IEEE CS Press, (34
28. Y. Wang, 2007aJuly). Software Engineering Foundations: A Software Science Perspective. CRC Book Series in Software Engineering, Vol. II, Auerbach Publications, NY.
29. B. Bouchon-Meunier, R. Yager, L. Zadeh, 1995Fuzzy Logic and SoftComputing. World Scientific, Singapore.
30. J. C. Bezdek, is. a. What, intelligence?. computational, Zurada. J. M. In, I. I. R. J. Marks, C. J. Robinson, eds.) Computational Intelligence: Imitating Life, 112IEEE Press, Los Alamitos (1994
31. L. A. Zadeh, Sets. Fuzzy, Information and Control 8, 338-353 (1965
32. Zadeh L.A.The roles of fuzzy logic and soft computing in the conception, design and deployment of intelligent systems. BT Technol J. 14199443236
33. R. A. Aliev, R. R. Aliev, Computing. Soft, I. I. I. I. I. I. volumes, A. S. O. A. Baku, Press, 1997-1998in Russian).
34. R. Aliev, K. Bonfig, F. Aliew, Computing. Soft, Verlag. Berlin, Technic, 2000
35. Zadeh L.A. Foreword. In Proc.First European Congress on Intelligent Techniques and Soft Computing- EUFIT'95, page VII, 1995
36. L. A. Zadeh, Soft Computing and Fuzzy Logic. IEEE Software 11199464858
37. R. A. Aliev, Expert. Fuzzy, In. Systems, F. Aminzadeh, M. Jamshidi, S. O. F. T. C. O. M. P. U. T. I. N. G. (eds, Logic. Fuzzy, Networks. Neural, Artificial. Distributed, 9910 . Intelligence.pages, N. , NJ: PTR Prentice Hall, 1994
38. Zadeh L.A. Fuzzy logic, Neural Networks and Soft Computing .Comm of ACM 37199437784
39. Welstead S.T. (ed) Neural Networks and Fuzzy Logic Applications in C/C++, Professional Computing.NY: John Wiley, 1994
40. Yager R.R. and Zadeh L.A. (eds) Fuzzy sets, neural networks and Soft Computing.NY: VAN Nostrand Reinhold , 1994
41. Nauck D., Klawonn F., and Kruse R., Foundations of Neuro-Fuzzy Systems.NY: John Wiley and Sons, 1997.
42. Mohamad H.Hassoun, Fundamentals of artificial neural networks.Cambridge: MIT Press, 1995
43. S. Haykin, Networks. A. Neural, Foundation. Comprehensive, Marmillau, I. E. E. E. Computer, Society, 1994
44. Goldberg D.E., Genetic algorithms in search, optimization and machine learning.Reading, MA: Addison-Wesley, 1989

45. Arciszewski, and De Jong K.A. (2001Evolutionary computation in civil engineering: research frontiers. Eight International Conference on Civil and Structural Engineering Computing, (Topping B. H. V., ed.), Saxe-Coburg Publications, Eisenstadt, Vienna, Austria.
46. K. Miettinen, P.and. Neittaanmaki, J. Periaux, 1999Evolutionary Algorithms in Engineering and Computer Science : Recent Advances in Genetic Algorithms, Evolution Strategies, Evolutionary Programming, John Wiley & Sons Ltd., pps. 483. 0-47199-902-4
47. R. R. Youngs, S. J. Chiou, W. J. Silva, J. R. Humphrey, 1997Strong ground motion attenuation relationships for subduction zone earthquakes. Seismol. Res. Lett., (68) 1, 58 EOF73 EOF
48. J. G. Anderson, 1997Nonparametric description of peak acceleration above a subduction thrust. Seismol. Res. Lett., (68) 1, 86 EOF93 EOF
49. C. B. Crouse, 1991Ground motion attenuation equations for earthquakes on the Cascadia subduction zone. Earth. Spectra, 7210236
50. S. K. Singh, M. Ordaz, M. Rodríguez, R. Quaas, V. Mena, M. Ottaviani, J. G. Anderson, D. Almora, 1989Analysis of near-source strong motion recordings along the Mexican subduction zone. Bull. Seism. Soc. Am., 7916971717
51. C. B. Crouse, Y. K. Vyas, B. A. Schell, 1988Ground motions from subduction-zone earthquakes. Bull. Seism. Soc. Am.,78125
52. S. K. Singh, E. Mena, R. Castro, C. Carmona, 1987Empirical prediction of ground motion in Mexico City from coastal earthquakes. Bull. Seism. Soc. Am., 7718621867
53. K. Sadigh, 1979Ground motion characteristics for earthquakes originating in subduction zones and in the western United States. Proc. Sixth Pan Amer. Conf., Lima, Peru.
54. B. F. Tichelaar, L. J. Ruff, 1993Depth of seismic coupling along subduction zones. J. Geophys. Res., 9820172037
55. G. M. Atkinson, D. M. Boore, 2003Empirical ground-motion Relations for Subduction-Zone Earthquakes and Their Applications to Cascadia and other regions. Bull. Seism. Soc. Am., 93417031729
56. S. C. Gómez, M. Ordaz, C. Tena, 2005Leyes de atenuación en desplazamiento y aceleración para el diseño sísmico de estructuras con aislamiento en la costa del Pacífico. Memorias del XV Congreso Nacional de Ingeniería Sísimica, México, Nov. A-II-02
57. D. Gasparini, E. H. Vanmarcke, 1976SIMQKE: A Program for Artificial Motion Generation, Department of Civil Engineering, Massachusetts Institute of Technology, Cambridge, MA.

58. W. J. Silva, K. Lee, 1987WES RASCAL code for synthesizing earthquake ground motions." State-of-the-Art for Assessing Earthquake Hazards in the United States, Report 24, U.S. Army Engineers Waterways Experiment Station, Misc. Paper S-731
59. B. A. Bolt, N. J. Gregor, 1993Synthesized Strong Ground Motions for the Seismic Condition Assessment of the Eastern Portion of the San Francisco Bay Bridge", Report UCB /EERC-93/12, University of California, Earthquake Engineering Research Center, Berkeley, CA.
60. J. E. y. C. A. Carballo, Cornell, 2000Probabilistic seismic demand analysis: spectrum matching and design", Department of Civil and Environmental Engineering, Stanford University, Report RMS-41
61. C. Kircher, 1993Personal communication with Farzad Naeim and Marshall Lew.
62. F. Naeim, J. Kelly, 1999Design of Seismic Isolated Structures from Theory to Practice. New York, John Wiley & Sons. 289p.
63. N. E. Huang, S. Zheng, S. R. Long, M. C. Wu, H. H. Shih, Q. Zheng, N. Yen, C. , C. C. Tung, M. H. Liu, 1998The empirical mode decomposition and Hilbert spectrum for nonlinear and nonstationary time series analysis", Proc. R. Soc. London, Ser. A 454903995
64. A. Roulle, F. J. Chavez-Garcia, 2006The strong ground motion in Mexico City: Analysis of data recorded by a 3D array, Soil. Dyn. Eq. Eng. 2671
65. H. Kawase, K. Aki, 1989A study of the response of a soft bas in for incident S, P and Rayleigh waves with special reference to the long duration observed in Mexico City, Bull Seism. Soc. Am. 7913611382
66. N. Yoshida, S. Iai, 1998Nonlinear site response and its evaluation and prediction," IN Irikura, K., Kudo, K., Okada, K. & Sasatani, T. (Eds.) The Second International Symposium on the Effects of Surface Geology on Seismic Motion, Yokohama, Japan, A.A.Balkema, 7190
67. I. y. Herrera, E. . Rosenblueth, Response Spectra on Stratified Soil". Proc. 3rd. World Conference on Earthquake Engineering. Nueva Zelandia, 44561965
68. M. P. Romo, A. Jaime, 1986Dynamic characteristics of some clays of the Mexico Valley and seismic response of the ground". Technical Report, Apr., Instituto de Ingenieria, Mexico City, Mexico (in Spanish).
69. A. Jaime, M. P. Romo, D. Reséndiz, 1988Comportamiento de pilotes de fricción en arcilla del valle de México." Series of the Instituto de Ingeniería, Mexico City, Mexico
70. M. P. Romo, Seed, 1986Analytical modelling of dynamic soil response in the Mexico Earthquake of September 19, 1985". Proc. ASCE Int. Conf. on the Mexico Earthquakes-1985148162

71. S. R. García, M. P. Romo, J. Mayoral, 2007Estimation of Peak Ground Accelerations for Mexican Subduction Zone Earthquakes using Neural Networks", Geofísica Internacional, 46-15163enero-marzo
72. G. Castro, S. J. Poulos, J. W. France, J. L. Enos, 1982Liquefaction induced by cyclic loading. Winchester, Mass: Geotechnical Engineers Inc.
73. H. B. Seed, I. M. Idriss, I. Arango, . Evaluation, Liquefaction. of, Using. Potential, Performance. Field, Data, of. Journal, Geotechnical. the, Division. A. S. C. E. Engineering, Vol, GT31983Seed et al., 1983
74. H. B. Seed, I. Idriss, M.1971Simplified procedure for evaluation soil liquefaction potential. Journal of the Soil Mechanics and Foundations ASCE, 97 (9), 1249-1273.
75. T. L. Youd, I. M. . Idriss, R. D. Andrus, I. Arango, G. Castro, J. T. Christian, R. Dobry, F. Liam, L. F. Harder, M. E. Hynes, K. Ishihara, J. P. Koester, S. S. C. Liao, I. I. I. W. F. Marcuson, G. R. Martin, J. K. Mitchell, Y. Moriwaki, M. S. Power, P. K. Robertson, R. B. Seed, K. H. Stokoe, 200, Liquefaction resistance of soils. Summary report from the 1996NCEER and 1998 NCEER/NSF workshops on evaluation of liquefaction resistance of soils. J. Geotech. Geoenviron. Eng., 127(10), 817-833.
76. R. Boulanger, I. M. Idriss, 2004State normalization of penetration resistance and the effect of overburden stress on liquefaction resistance. Proc. 11th International Conf. on Soil Dynamics and Earthquake Engineering and 3rd International Conference on Earthquake Geotechnical Engineering, Univ. of California, Berkeley, CA.
77. S. M. Fatemi-Agdha, M. Teshnehlab, A. Suzuki, T. Akiyoshi, Y. Kitazono, 1998Liquefaction potential assesment using multilayer artificial neural network. J. Sci. I.R. Iran, 9(3).
78. C. H. Juang, C. J. Chen, Y. M. Tien, 1999Appraising cone penetration test based liquefaction resistance evaluation methods: Artificial neural networks approach. Canadian Geotechnical Journal, 363443454
79. C. H. Juang, H. M. Yuan, D. H. Lee, Lin, 2003P. S., "Simplified cone penetration test-based method for evaluating liquefaction resistance of soils," Journal of Geotechnical and Geoenvironmental Engineering, 12916680
80. M. H. Baziar, N. Nilipour, 2003Evaluation of liquefaction potential using neural-networks and CPT results. Soil Dynamics and Earthquake Engineering, 237631636
81. S. Chern, C. Lee, 2009CPT-based simplified liquefaction assessment by using fuzzy-neural network. Journal of Marine Science and Technology, 174326331
82. García S.R. and Romo M.P.2007GENES: Genetic Generator of Signals, a Synthetic Accelerograms Application. , Proc. of the SEE5, SM-80.

83. S. F. Bartlett, T. L. Youd, 1992Empirical Analysis of Horizontal Ground Displacement Generated by Liquefaction Induced Lateral Spreads", Tech. Rept. NCEER 920021National Center for Earthquake Engineering Research, SUNY- Buffalo, Buffalo, NY.
84. Youd T.L., Hansen C.M. and Bartlett S.F.2002Revised Multilinear Regression Equations for Prediction of Lateral Spread Displacement' Journal of Geotechnical and Geoenvironmental Engineering, 1281210071017

CITATION

Silvia Garcia (2012). A Cognitive Look at Geotechnical Earthquake Engineering: Understanding the Multidimensionality of the Phenomena, Earthquake Engineering, Prof. Halil Sezen (Ed.), ISBN: 978-953-51-0694-4, InTech, DOI: 10.5772/50369.

CHAPTER 2

Seismic Performance of Historical and Monumental Structures

Halil Sezen[1] and Adem Dogangun[2]

[1]Department of Civil, Environmental, and Geodetic Engineering, The Ohio State University, Columbus, Ohio, USA
[2]Department of Civil Engineering, Uludag University, Bursa, Turkey

INTRODUCTION

Historical, religious and monumental structures and their susceptibility to damage in recent earthqaukes in Turkey are presented and discussed in this chapter. Turkey has a very large number of historical structures and is located in one of the most seismically active regions of the world. Some of these historical and monumental masonry and reinforced concrete structures suffered substantial damage or collapsed during two major earthquakes in 1999. The Kocaeli (M_w7.4) and Düzce (M_w7.2) earthquakes occurred on August 17 and November 12, 1999, and ruptured approximately 110 km and 40 km of the 1550-km-long North Anatolian fault, respectively. This chapter describes briefly the construction materials and techniques for historical religious and monumental structures and state-of-practice in Turkey, and presents dynamic analyses of a masonry minaret example. The seismic performance of the mosques and minarets (tall slender towers) during the 1999 earthquakes is presented.

SEISMIC DESIGN AND CONSTRUCTION PRACTICE IN TURKEY

Prior to the 1999 earthquakes, two codes governed the design and construction of reinforced concrete and masonry buildings in Turkey: Turkish Earthquake Code (1998) and the Turkish Building Code, TS-500 (1985). The earthquake code included procedures for calculating earthquake loads on buildings. The ductility requirements and details described in the earthquake code were rarely observed in religious or monumental structures inspected by the authors after the 1999 earthquakes. Details of the seismic design and building construction practice prior to the 1999 earthquakes are provided in Sezen et al. (2003).

The descriptions of ground motion characteristics, structural damage, and performance of structures during these earthquakes are provided in Sezen et al. (2003). Response spectra for selected acceleration histories for 5% damping are presented in Figure 1. The ground motions included in Figure 1 were recorded at: SKR station in Adapazari (Peak Ground Acceleration, PGA 0.41g, stiff soil); YPT station in Yarimca (PGA 0.23g, soft soil); and DZC station in Düzce (PGAs: EW-Aug.17 0.36g, NS-Aug.17 0.31g, EW-Nov.12 0.54g, and NS-Nov.12 0.35g, soft soil). Figure 1 also provides a comparison between the linear elastic acceleration response spectra calculated for rock and that calculated for soft soil sites using the provisions of the Uniform Building Code (UBC, 1997) and those of the Turkish seismic code (1998) with 5 percent damping for the highest seismicity in the United States and in Turkey, respectively. From Figure 1, it can be concluded that, for a structure with given periods of vibration, the difference between the base shear calculated using the 1998 Turkish seismic code and the base shear demand obtained from the recorded acceleration response spectra do not differ significantly. If the static lateral load distribution over the height is suggested by the code is assumed to be credible, structures designed and detailed according to the Turkish code should not have collapsed or suffered severe damage during the 1999 earthquakes.

Specifically, according to Figure 1, a structure with a fundamental period of 0.5 seconds would be subjected to seismic forces larger than those specified in the code. The period of older masonry structures with thick and shorter walls tends to be relatively small, probably 0.5 seconds or less. On the other hand, the minarets or very slender towers tend to be very flexible with relatively large fundamental periods. Figure 1 shows that the recorded spectral accelerations are significantly large at large periods. The authors‘ field observations after the 1999 earthquakes, especially the November 12 Duzce earthquake (DZC), showed more damage in slender structures like minarets. In addition, among other factors leading to inadequate performance, the extent of damage observed in most reinforced concrete structures including minarets after the 1999 earthquakes is probably related to poor engineering and lack of conformance to the relatively new Turkish seismic code.

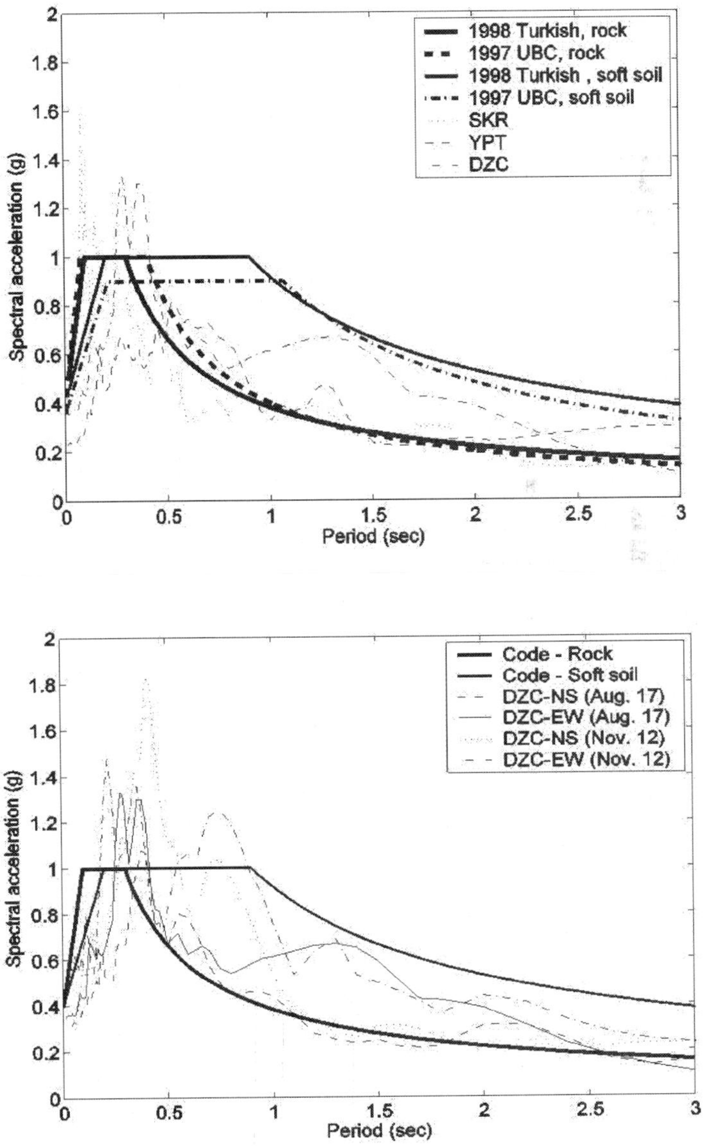

Figure 1. Absolute Acceleration Response Spectra

Construction Materials and Deterioration of Historical Structures

In the region affected by the earthquakes, in general, construction materials include stone or brick masonry and wood for older mosques and minarets, steel and reinforced concrete in more recently constructed structures. Typically locally available stone blocks were used. A special mortar called Horasan mortar was used in historical masonry structures. The Horasan mortar is made by mixing lime with baked clay powder that could be obtained by grinding clay tiles or bricks. The Horasan mortar was commonly used in the Middle East, where use of earthenware, pottery and brick construction were widespread (Çamlıbel 1988).

Hollow clay tiles or lightweight concrete blocks are used as infill wall material in recently built reinforced concrete (RC) moment frame mosques. Solid clay bricks were used in infill walls of RC mosques built before early 1970s. Almost all historical religious structures in Turkey were constructed using cut stone, masonry blocks or combination of these two materials. Timber construction is rare and reinforced concrete is relatively recent. The structural and geometrical properties of each masonry structure depend on many factors including the structural knowledge and applications at the time of construction, experience of the architect or engineer, seismicity of the region, and availability of construction materials in that area. Recent earthquakes in Turkey have shown that most masonry monumental structures as well as buildings (Ural et al. 2012) in high seismic regions are vulnerable to structural damage and collapse.

Historical structures deteriorate over time mainly due to environmental effects, and therefore may experience failure or collapse under gravity loads or seismic loads lower than those predicted by the design codes. As discussed in Dogangun and Sezen (2012) and Sezen and Dogangun (2009), in many cases, the following factors trigger or exacerbate the damage: (1) Surface or rain water runoff. If the structure and drainage system are not maintained properly, grass or fungus may grow and weaken the structural materials. Water accumulated on or penetrated into structural members may cause cracks due to freezing and thawing. (2) Soil settlement and relative movement of foundation. (3) Insufficient material strength. Layers of clay or other impure materials inside stone blocks may eventually lead to wearing, spalling or cracking. In stone masonry structures, the properties of the mortar significantly influence the strength of the entire structural component such as a load bearing wall. Deterioration of mortar binding the stone blocks, especially poor quality mortar including mud or low quality lime, can reduce the strength and stiffness of the wall considerably. (4) Other problems. Historical structures can be subjected to various environmental and loading conditions depending on their use and geographic location. For example, timber is more susceptible to humidity and temperature variations. Loading from continuous traffic and heavy trucks can lead to vibrations and excessive loads on foundations because the streets and other structures in historical cities are not designed for modern day traffic. Similarly, the use or occupancy of the structure may change and create larger unexpected loads. Parts of older structures are sometimes used as storage, in which the magnitudes of loads are usually much higher. Other local and environmental effects, such as acid rains, may adversely affect construction materials.

Most of the factors presented here result in gradual deterioration of materials or the load carrying structural system, which can be prevented as the damage progresses and becomes visible in many cases. On the other hand, structural damage, failure or total structural collapse occurs suddenly during moderate or strong earthquakes. Thus, it is essential to evaluate the capacity of existing historical structures and to retrofit them before an expected earthquake strikes.

Construction Practice and Techniques

Overall plan dimension of a typical mosque generally varies between 12 and 25 meters. Depending on the plan dimensions, height ranges from 7 m to 15 m not including the dome. Height of the dome, a half sphere, is typically half of the plan dimension (Figure 2). Lateral seismic loads are typically resisted by relatively thick unreinforced stone masonry walls in historical mosques. Addition to load bearing masonry walls, most mosques include a few columns typically carrying the gravity loads.

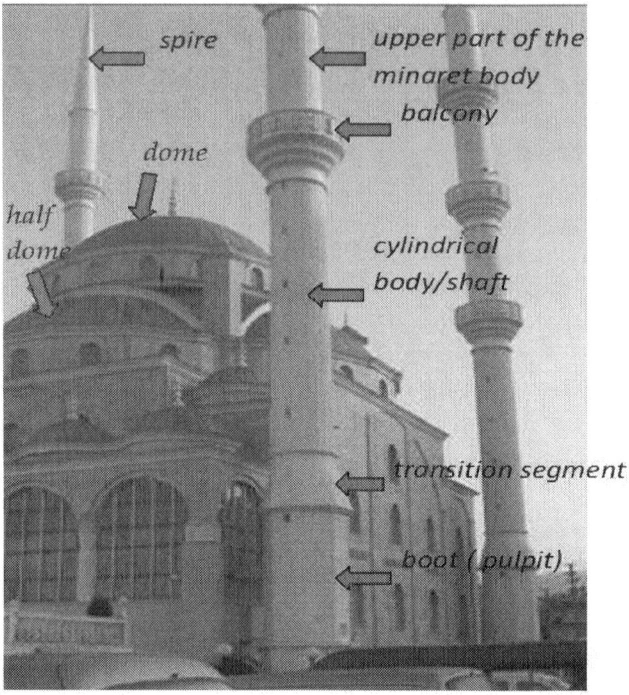

Figure 2. Typical mosque and its minarets in Turkey

Minarets can be separate or contiguous and integral with the mosque structure, and are typically built using stone, brick, wood or reinfroced concrete. They typically include cylindrical or polygonal body/shafts, one or two balconies, and a conical roof or spire (Figure 2). In masonry minarets or slender tower structures, rather small tensile strength of mortar placed between the masonry blocks presents a major problem in regions of high seismicity. The brick or stone blocks have fairly large compressive strength, however unreinforced masonry lacks tensile strength required to resist bending moments imposed by the lateral earthquake loads. Older masonry minarets were typically constructed using stone blocks or solid clay bricks or a combination of two, whereas unreinforced lightweight stone blocks are preferred in new construction.

After a major earthquake in 1509, Ottoman architects tackled the problem of constructing tall earthquake resistant minarets (Oğuzmert, 2002). They started to use a special technique for linking adjacent stone blocks with iron bars and clamps in the vertical and horizontal directions as shown inFigure 3 (Doğangün et al. 2007). Use of iron clamps in the two perpendicular directions (transverse and vertical) has improved the lateral load carrying capacity of slender masonry minarets significantly under earthquake loads. The clamps and vertical bars were placed inside anchorage holes in the stone blocks, and melted lead was poured inside the hole to provide bond between the stone and iron clamp or vertical bars. Depending on the properties and dimensions of the stone units, different clamps were developed. For example, as shown in Figure 4a, curved clamps were used within the circular stone wall on the minaret perimeter. Sharper and thinner clamps in thin stone blocks (Figure 4b) and shorter clamps were used if low tensile stresses are expected (Figure 4c). Approximately 2000 kilograms of this heavy metal, lead was used for the construction of a typical masonry minaret. Lead performs as intended for a very long time because it does not corrode or is hardly ever influenced by the adverse environmental conditions.

POST-EARTHQUAKE SURVEYS

The data presented here are based on observations from two surveys conducted after the August 17 and November 12, 1999 earthquakes. Damage to historical and recently constructed mosques and minarets was documented to investigate the seismic performance of these structures. The main objectives of these surveys were to provide detailed information about the characteristics of the observed damage, and to study the relative vulnerability of these historical and modern structures to strong ground motions. The parameters considered in the survey included: type of construction material, minaret height, location and description of damage, and location (coordinates, if available). Vast majority of the surveyed minarets and mosques were located in the cities of Düzce and Bolu. The peak ground accelerations recorded in Düzce during both earthquakes were larger than 0.30g (Figure 1), whereas that recorded in Bolu was reported as 0.82g after the November 12, 1999 earthquake.

Structural performance levels and corresponding representative sample damage descriptions for mosques and minarets are shown in Table 1. It should be noted that non-structural damage is irrelevant for minarets. In general, damage to the non-structural components of the mosques was insignificant as compared to the total structural damage.

A total of 59 sites were visited after the October 12, 1999 Duzce earthquake (second earthquake). The name, location, construction date, and observed damage levels are provided in Tables 2 and 3. The first 22 mosques listed in Table 2 were located in the city of Duzce and neighboring town of Kaynasli. Mosques numbered 23 through 44 were located in the city of Bolu.

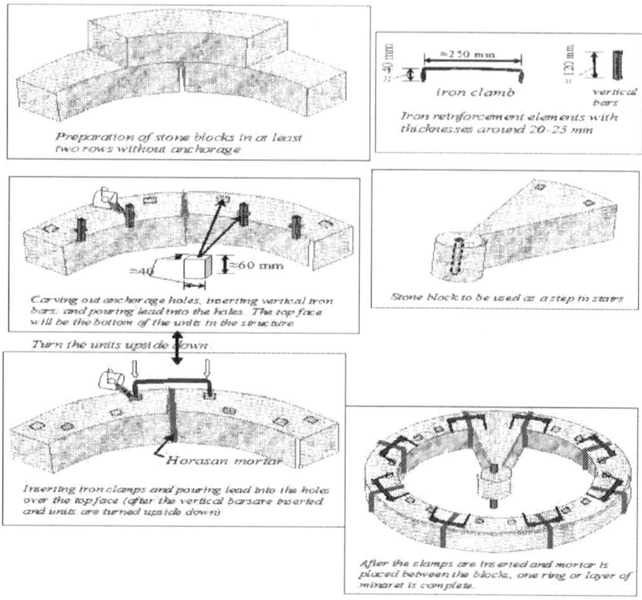

Figure 3. Construction of traditional Turkish minarets using stone blocks reinforced and anchored with iron bars and clamps (Dogangun et al. 2007)

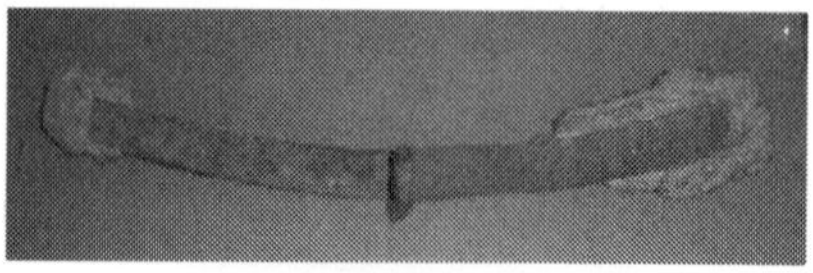

(a)

(b)

(c)

Figure 4. Variations in iron clamps used in historical stone masonry walls

Table 1. Structural damage description and classification for mosques and minarets

Performance level	Damage classification	Sample damage description (mosque)	Sample damage description (minaret)
I	None	Negligible	Negligible
II	Light	Minor cracks in primary structural components	Minor cracks in masonry or RC minarets
III	Moderate	Significant cracks in RC members or masonry walls	Significant cracks especially around the minaret base
IV	Major	Hinge formation and wide cracks in primary RC members Infill wall collapse	Permanent visible drift
			Wide cracks and concrete spalling
V	Collapse	Partial or total collapse	Collapse

Table 2. Damage to the minarets and mosques surveyed in Duzce and Bolu

Name	Location (coordinates)		Construction date	Damage level	
	North	East		Mosque	Minaret
1.Sirali Koyu	40°49.246'	31°11.544'	1987	Light	Collapse
2.Kocyazi Koyu	40°50.569'	31°10.249'	1977-79	Light	Collapse
3.Karaca	40°50.733'	31°09.950'	1970s	Collapse	Collapse
4.Hamidiye	40°50.838'	31°09.518'	1980s	Light	Light
5.Rumelipalas	40°50.877'	31°08.444'	1972	None	None
6.Uzun Mustafa	40°50.755'	31°08.736'	1989	None	Collapse
7.Kultur Mahallesi	40°50.623'	31°09.257'	-	None	Moderate
8.Otopark	40°50.446'	31°09.119'	1971-73	None	Moderate
9.Aydinpinar	40°49.796'	31°09.190'	1994	Light	Collapse
10.Yesil	40°49.966'	31°09.480'	1990	None	Collapse
11.Asar	40°50.012'	31°09.650'	1977	None	Moderate
12.AzmimilliYeni	40°49.897'	31°10.087'	1988	None	Collapse
13.Mimar Sinan Nur	40°50.006'	31°10.236'	1988-91	None	Collapse
14.Maresal F.Cakmak	40°49.928'	31°10.525'	1990	Light	Collapse

15.Huzur	40°49.680’	31°11.324’	1986	None	None
16.Topalakli	40°49.606’	31°11.539’	1951	Light	None
17.Kirazli Koyu	40°48.918’	31°12.844’	1965	None	None
18.Doganli Koyu	40°48.182’	31°14.190’	1981-94	None	None
19.Uckopru Merkez	40°47.662’	31°15.018’	1985	None	Light
20.Yesiltepe	40°46.690’	31°17.467’	1986-90	Collapse	Collapse
21.Karacaali	40°46.496’	31°18.204’	1992-94	None	Moderate
22.Kaynasli merkez	40°46.406’	31°19.138’	-	Collapse	Collapse
23.Sanayi	40°44.280’	31°37.562’	1988	None	None
24.Kultur	40°44.507’	31°36.340’	1990	None	Light
25.Oksuztekke	40°44.488’	31°35.851’	1993	Major	Collapse
26.Ozayan	40°44.638’	31°35.405’	1996	None	Moderate
27.Beskonaklar Yeni	40°44.263’	31°35.599’	1997	Moderate	-
28.Pasakoy Berberler	40°43.942’	31°34.096’	1987	None	None
29.Pasakoy Eniste	40°43.803’	31°34.689’	1997	None	Light
30.Sumer	40°43.575’	31°35.587’	1992-95	None	Light
31.Sumer Mah. Yeni	40°43.468’	31°35.297’	1983	None	None
32.Karacayir Siteler	40°43.577’	31°36.033’	1998-99	None	-
33.Semsi AhmetPasa	40°43.852’	31°36.635’	14th cent	Major	Collapse
34.Sarachane	40°43.935’	31°36.513’	17th cent	Light	None
35.Yildirim Bayezid	40°44.040’	31°36.576’	1804	Major	None
36.Kadi	40°43.901’	31°36.459’	1499	Major	Collapse
37.Aslahaddin	40°43.981’	31°36.732’	1978	None	None
38.Balci	40°43.885’	31°37.111’	1990	None	None
39.Aktas	40°43.783’	31°36.650’	1900s	Major	None
40.Karacayir	40°43.825’	31°36.515’	1946	None	None
41.Kabaklar	40°44.859’	31°36.111’	1981	None	None
42.Camli	40°44.243’	31°35.841’	1980s	Major	None
43.Sultanzade	40°44.249’	31°35.817’	1930s	Major	Major
44.Yesil	40°44.170’	31°36.170’	1966	Major	None

Table 3. Damage to the other mosques and minarets visited (coordinates not available)

No.	Name	Location	Mosque	Minaret
45	Cumhuriyet Mah.	Duzce, downtown	None	Collapse
46	Merkez	Duzce, downtown	Major	Collapse
47	Cedidiye Merkez	Duzce, downtown	Light	Collapse
48	Yuvacik	Yuvacik village, near Golcuk	Major	-
49	Asagi Yuvacik	Yuvacik village, near Golcuk	None	None
50	Yeni	Adapazari	None	Collapse
51	Yalova	Yalova, downtown	Light	None
52	Izmit (1)	Izmit, next to highway E5	Light	Collapse
53	Izmit (2)	Izmit, next to highway TEM	None	Light
54	Izmit (3)	Izmit, downtown	Major	Light
55	Golyaka	Golyaka, downtown	Major	Collapse
56	Suleymanbey	5 km east of Golyaka	Collapse	-
57	Golcuk (1)	4 km west of downtown	Major	-
58	Golcuk (2)	Near Ford plant, west of city	Collapse	Moderate
59	Dariyeri Hasanbeyi	East of Duzce (a village)	Major	Light

Before mid-1960s, major construction materials were wood and stone or brick masonry. Most of the recently constructed mosques and minarets were reinforced concrete. All of the reinforced concrete mosques were built after 1965. In older mosques, solid bricks were used in infill walls. Hollow clay tiles were used as infill material in the mosques built after late 1970s. As shown in Figure 5, 84 percent of the minarets surveyed in Duzce and Kaynasli were reinforced concrete, whereas only 46 percent of the minarets were reinforced concrete in Bolu. Unreinforced stone masonry was commonly used in old minarets as well as in the minarets constructed in recent years.

OBSERVED EARTHQUAKE DAMAGE

The Figure 6 shows the damage distribution for the mosques and minarets surveyed. Damage distribution for RC mosques and minarets are presented in Figures 6c and 6d separately.

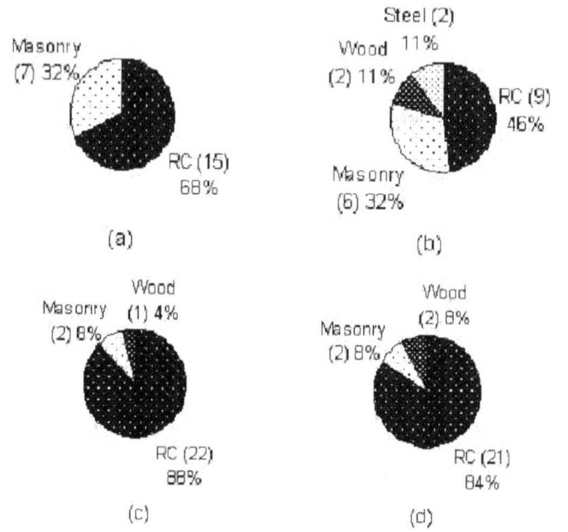

Figure 5. Type of construction for (a) mosques, (b) minarets in Bolu; and (c) mosques, and (d) minarets in Duzce and Kaynasli (number of mosques/minarets is given in parenthesis)

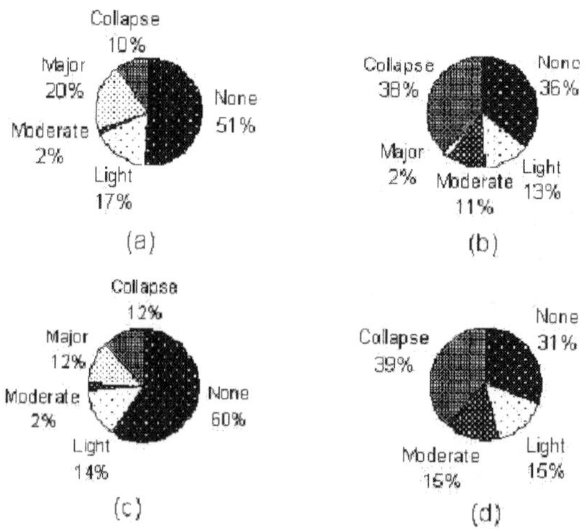

Figure 6. Damage distribution for (a) all mosques and (b) all minarets, and damage distribution for (c) RC mosques and (d) RC minarets

Comparison of Figures 6a and 6c indicates that the damage in RC mosques was less as compared to other structural systems. More than ten percent of the mosques surveyed collapsed. Three of the 26 mosques surveyed in

Duzce and Kaynasli region collapsed. There was no collapsed mosque in Bolu, but five of the 21 mosques surveyed had to be closed after thc October 12, 1999 earthquake. Closed Mosques in Bolu were Semsi Ahmet Pasa (Imaret), Yildirim Bayezid, Kadi, Camli and Yesil. Note that the first three of these mosques are at least 200 years old.

As illustrated in Figures 6b and 6d, percentage of all damaged minarets and only RC minarets are similar. Almost forty percent of the minarets collapsed, and approximately one third of the minarets were undamaged. Failure plane for almost all collapsed RC minarets was within 1.5 meter long region above the minaret base or pyramid-shaped transition segment (Figure 7a), where the longitudinal reinforcing bars were usually spliced. Horizontal circumferential cracks and spalling of concrete were commonly observed at the bottom of the cylindrical body of RC minarets (Figure 7a). Frequently, flexural cracks, concrete crushing or spalling was observed in this region of the damaged RC minarets. As shown in Figure 7, less frequently collapse or damage occurred within the transition segment and middle of the cylindrical body or near the top of the minaret.

(a) (b) (c)

Figure 7. Collapse of a minaret near the bottom of cylinder (left), minaret damage within transition segment (middle), and collapsed minaret at mid-height of cylinder (right)

The ratio of collapsed or damaged unreinforced solid brick or stone masonry minarets was much larger than that of RC minarets. As listed in Table 4, majority of the visited masonry minarets collapsed (Sezen et al. 2003, and Firat 1999). Note that most of these minarets were either very new or few hundred years old (Tables 2 and 3). A minaret may either have an independent foundation (referred to as Type I minaret hereafter) or the base of the minaret may be attached to the roof of the mosque (referred to as Type II minaret). The minarets in Figures 2 and 7 are Type I and II minarets, respectively. In Type II minarets, the minaret and mosque structure respond independently. Large deformatiosn or failure in one structure does not affect the other one.

Lateral laod resisting mechanism of minarets are quite different from that of other structures. The height of center of minaret's mass can be very high above ground, resulting in large bending moments and shear forces. Masonry minarets without reinforcement or clamps (Figures 3 and 4) and with weak mortat are the most vulnerable against earthquakes. Masonry minarets typically either collapsed or suffered minor or no damage. This suggests that these minarets had no ductility or very limited deformation capacity. If the demand due to lateral earthquake forces is less than minaret's ultimate strength, minaret behaves elastically and suffers no damage. However, as soon as lateral demand exceeds the elastic capacity, the minaret collapses. Unlike RC minarets, masonry minarets failed at different locations along the height of the minaret inconsistently (Figure 8). The Imaret and Kadi mosques and possibly their minarets were 600 and 500 year old, respectively. As shown in Figures 8aand 8b, their minarets collapsed at the bottom of the cylindrical body. On the other hand, the Oksuztekke minaret collapsed at its mid-height (Figure 8c). Oksuztekke mosque was constructed only six years before the 1999 earthquakes (Table 2).

Table 4. Masonry minarets surveyed after the 1999 earthquakes

Name	City	Location or coordinates	Type	Observed damage
Aziziye Merkez	Düzce	40.50N-31.08E	II	Failed at the bottom of cylinders and collapsed
Cedidiye Merkez	Düzce	40.50N-31.09E	I	Failed at the bottom of cylinders and collapsed
Oksuztekke	Bolu	40.44.488N-31.35.851E	I	Minaret segment above the 2nd balcony level collapsed
Semsi Ahmet Pasa (Imaret)	Bolu	40.43.852N-31.36.635E	I	Failed near the bottom of the cylinder and collapsed
Sarachane	Bolu	40.43.935N-31.36.513E	I	Cracks in stone blocks in the mosque – no observed minaret damage
Yildirim Bayezid	Bolu	40.44.040N-31.36.576E	II	Dislocation of stone blocks in the mosque – no obserevd damage in minarets
Kadi	Bolu	40.43.901N-31.36.495E	I	Severe damage to mosque, minaret collapsed

Earthquake Damage in Historical Mosques

Among the historical mosques surveyed, Düzce Merkez, 600-year-old Imaret, 500-year-old Kadi, 300-year-old Sarachane, and 200-year-old Yildirim Bayezid mosques observed to suffer significant structural damage after the 1999 earthquakes (Dognagun and Sezen, 2012). The Yıldırım Bayezid mosque was originally built in 1382 and was burned down in the 19th century. A new structure was constructed after the fire, and it was severely damaged during a 7.3 magnitude earthquake in 1944. The mosque was damaged during the 1999 earthquakes and was closed for a period of time. On the south side of the mosque, portion of the walls above and below the windows were subjected to larger shear stresses (compared to solid wall sections) during the strong ground shaking. Higher shear demand in those parts of the relatively thick walls created serious cracks and openings between the stone blocks (Figure 9). As shown in Figure 9 (middle photo), portion of the walls above and below the windows act, in a sense, like short columns during the earthquake. Larger shear demand in those parts of the relatively thick walls created wide cracks between the stones.

(a) (b) (c)

Figure 8. Failures near the bottom and mid-height of the masonry cylindrical minaret body: (a) Imaret, (b) Kadi, and c) Oksuztekke mosques in Bolu

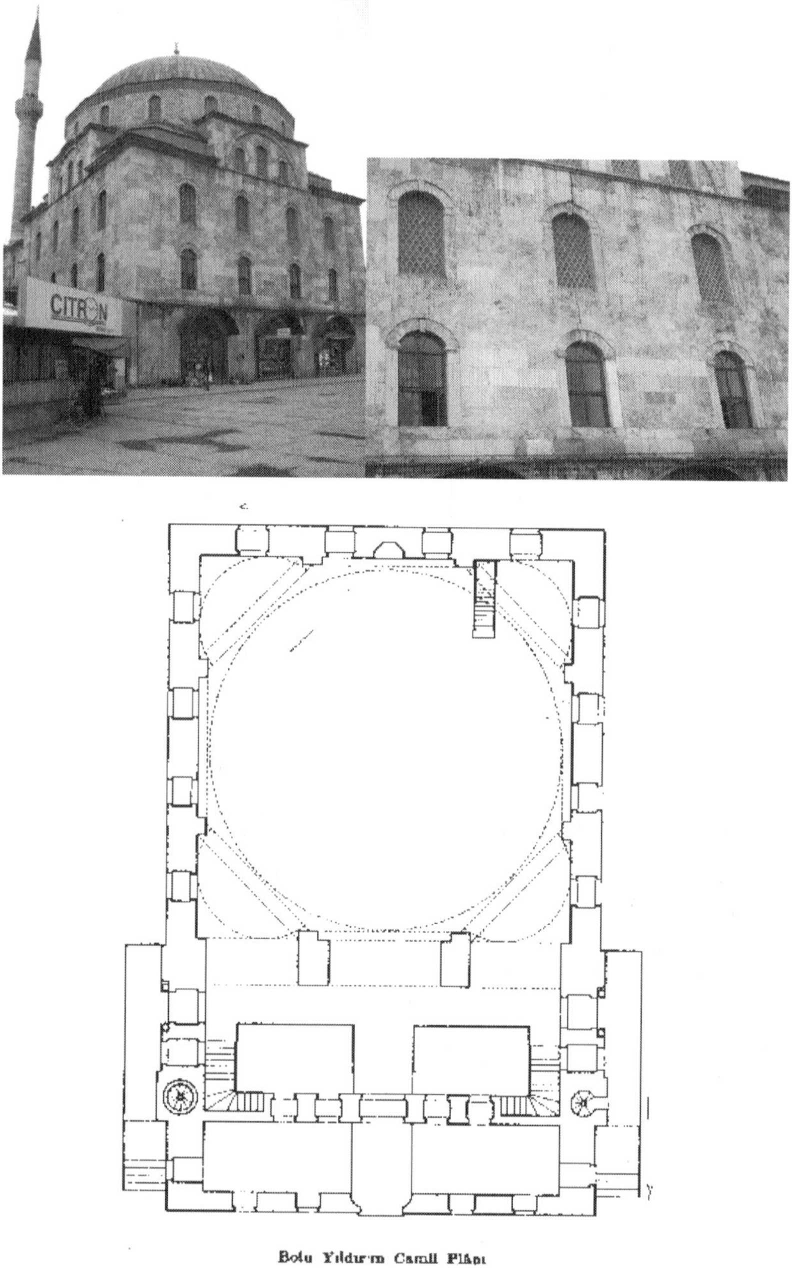

Figure 9. Wall damage and plan view of historical Yildirim Bayezid mosque in Bolu

Another historical mosque, Kadi (Figures 8b and 10a) sustained substantial damage including large cracks around historical entrance door, severe cracks and stone dislocations in the 1.5 m wide unreinforced stone masonry perimeter walls. Damage was mostly concentrated below or above the windows. The missing two key keystones and dislocated stones on top of the upper window are visible in Figure 10a. The reduced wall area along the vertical section through the windows was stressed more, causing considerable damage in those wall areas. The stone masonry minaret collapsed right above its base because the minaret base was integral with minaret walls (Figure 8b) and was quite stiff compared to the cylindrical minaret body.

The Imaret mosque, one of the oldest structures in the region, and its minaret were built using stones and small solid bricks bounded by a thick layer of mortar. The mortar between the bricks is typically as thick as the bricks. Old brick masonry minaret collapsed and Imaret mosque was closed after the 1999 earthquakes due to cracks in the walls (Sezen et al. 2003). The 1999 earthquakes caused very limited damage to the 300-year-old Sarachane mosque and there was no visible damage to its minaret. Some large cracks were observed in the walls at locations similar to those observed in other mosques discussed in this paper. Figure 10b and 10c show examples of couple such cracks; one immediately above a window and another in a corner near the roof.

(a) (b) (c)

Figure 10. a) cracks developed in the walls and above windows of Kadi; and b, c) damage in in the walls of Sarachane mosque

DYNAMIC ANALYSIS OF REPRESENTATIVE MASONRY MINARETS

Although numerous collapses and structural damage to minarets and mosques were documented after strong earthquakes, only few researchers investigated the seismic behavior and performance of minarets (Sezen et al. 2008, Portioli et al. 2011, Turk and Cosgun 2012, Altunisik 2012, Oliveira et al. 2012). It is essential to understand the dynamic behavior of these structures to improve the life safety and to preserve and strengthen the historical monumental structures. In this research, as an example, dynamic modeling and seismic analysis of unreinforced masonry minarets is presented.

Recently, a study was conducted by the authors to investigate the seismic response of reinforced concrete (RC) and masonry minarets (Dogangun et al. 2008, and Sezen et al. 2008). Modal analysis and dynamic time history analyses were conducted. Strength and deformation capacities of the RC and unreinforced masonry models are influenced by unique differences in material properties, including weight, modulus elasticity, and Poisson's ratio. For example, the flexural strength of an unreinforced masonry model is significantly lower as the reinforcing steel in a similar RC model provides tensile resistance. While Sezen et al. 2008 investigates distinct parameters affecting the behavior of RC minarets using a single model, the example presented below examines the effect of masonry material properties and minaret height on the minaret response using three different models. Detailed modeling properties and analysis results are presented in Dogangun et al. (2008).

Although the height of a minaret varies greatly depending on many factors such as the location or construction materials used, the height of a typical minaret with a single balcony over its height is about 20 to 25 m. In this research, two generic minarets with a height of 25 and 30 m and double balconies (Minaret I and Minaret II) and another 20 m tall minaret with a single balcony (Minaret III) were modeled and analyzed. As shown in Figures 11 and 12, only the height of the cylindrical minaret body was varied while the geometrical properties of the base or boot, transition segment, balconies and conical cap were kept the same in all three models. It should be noted that not all conical caps at the top of the masonry minarets are constructed using masonry materials. For uniformity, all minarets including the conical caps are modeled with the same masonry material in this study.

The finite element models of the three minarets are shown in Figure 12, and significant mode shapes of a typical minaret are shown in Figure 13 (Sezen et al. 2008). The computer program, ANSYS was used to model and analyze the minarets (see Dogangun et al. 2008 for details). In the models, the modulus of elasticity of uncracked section, Poisson's ratio and unit weight of masonry material (ordinary limestone) are taken as 3000 MPa, 0.2, and 20 kN/m^3, respectively. In the models, linear elastic material models and five percent damping ratio is used in all dynamic time history analyses. It is assumed that the minaret is located in a high seismic region Zone 1 in Turkish Earthquake Code (TEC, 2007). Considering the TEC, a structural behavior factor, R of 3 and an importance factor, I of 1.2 can be used for such structures.

Dynamic time history analyses are carried out for each minaret model using two ground motions recorded during the 12 November, 1999 Duzce and 17 August, 1999 Kocaeli earthquakes (see Dogangun et al. 2008 for details). The input ground motion was applied only in one horizontal direction. The modal periods of the minaret models (calculated from the modal analysis) and their contribution to the total dynamic response are calculated. The modal periods are greatly affected by the height of the minaret which affects the total mass and flexural stiffness. The calculated first or fundamental periods were 1.41, 0.90, and 0.51 seconds for the minaret models I, II, and III, respectively. The calculated modal response quantities indicate that the contribution of higher mode effects to total dynamic response is significant. The first mode contribution to the total response is about 47 or 49 percent. The torsional or the fifth mode has virtually no effect on the total response of the almost symmetrical minaret structures.

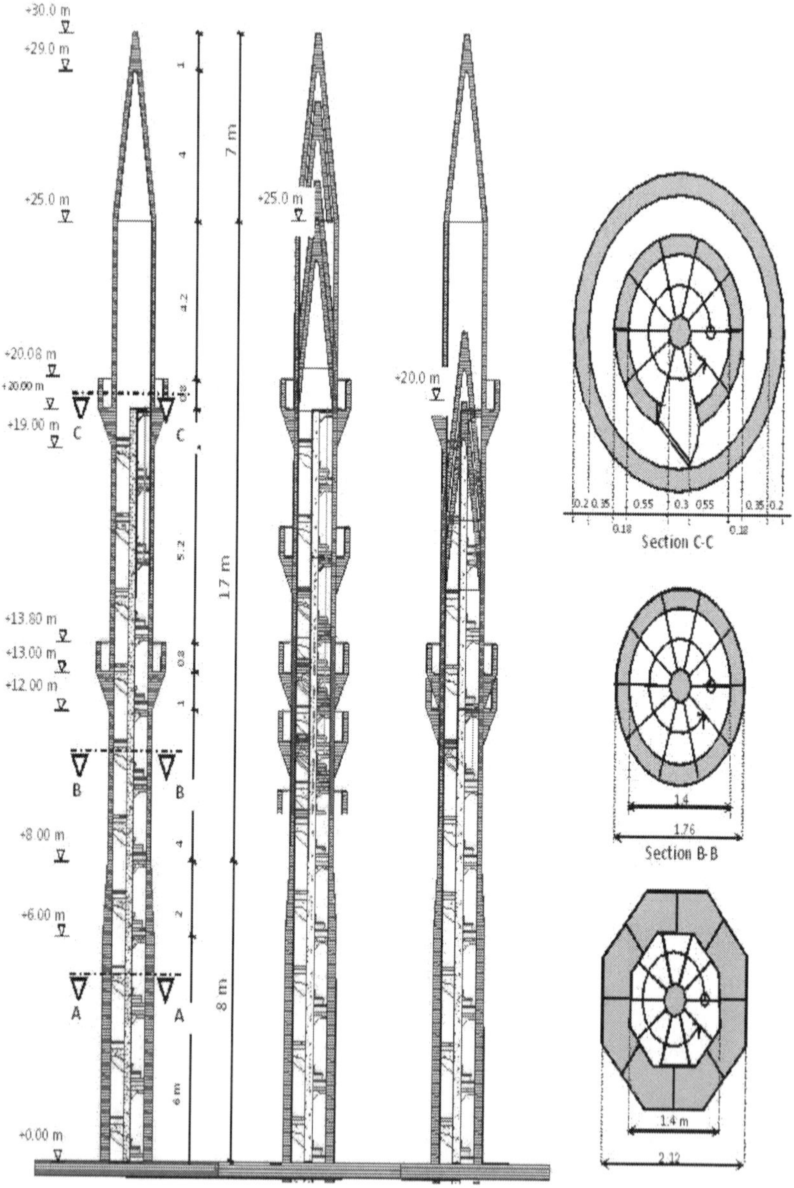

Figure 11. Three minaret models and their geometrical and cross-sectional properties

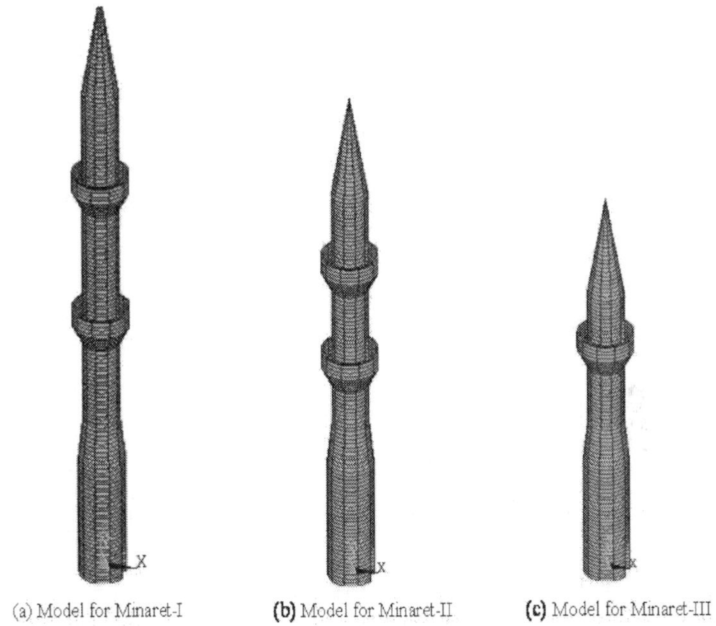

Figure 12. Finite element models of the three minarets

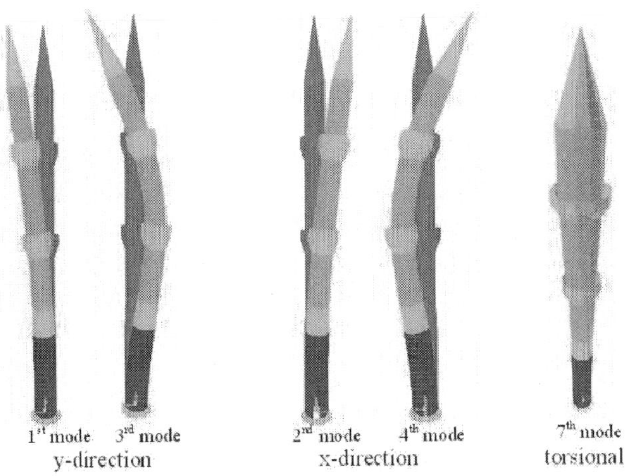

Figure 13. First two translational and the seventh mode shapes for a typical model

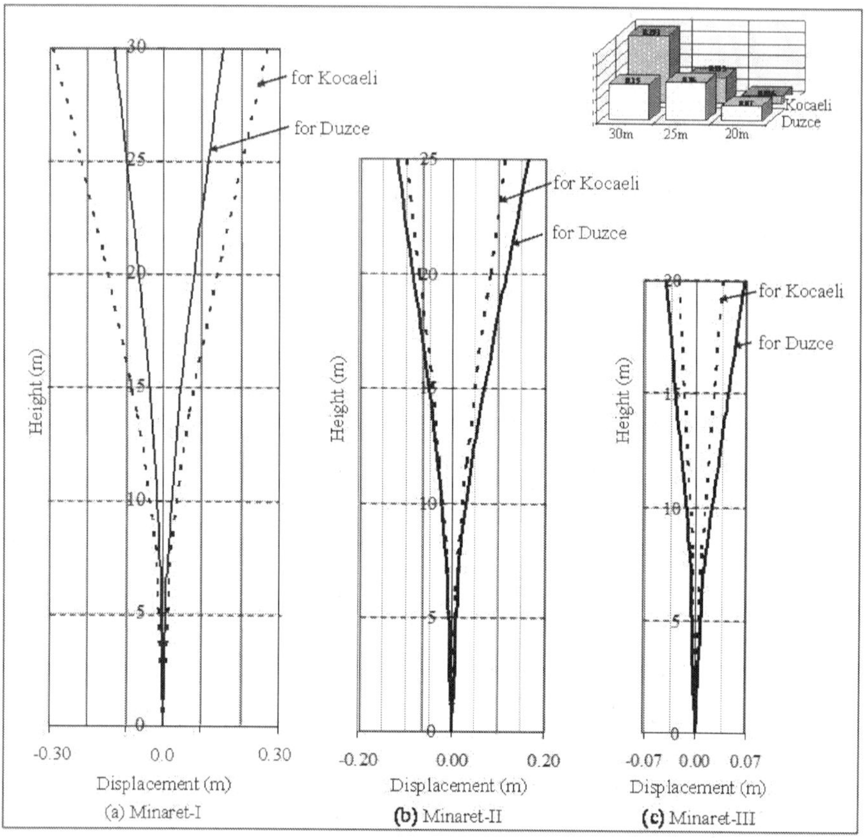

Figure 14. Lateral displacement distribution over the height of minarets for Duzce and Kocaeli ground motions

Figure 14 shows the calculated maximum elastic lateral displacement distribution along the length of three minaret models subjected to 1999 Duzce and Kocaeli earthquake ground motions. As shown inFigure 14 the deflected shapes of the minarets appear to show flexure dominated response with the largest displacements calculated at the top. Although the minarets act as cantilevers, the deformations are much smaller over the height of relatively stiff 6 m high base or boot. The displacements start to increase above the transition segment at about 8 m from ground level. The maximum dynamic displacement of Minaret I is 88% larger for the Kocaeli input motion (0.291 m) than that for the Duzce ground motion (0.154 m), although the maximum acceleration of the Duzce motion was larger. For the other two shorter minarets Minaret I and II, the maximum calculated displacements are larger for the Duzce ground motion. The maximum dynamic lateral displacement of 0.29 m calculated for the 30 m tall minaret under Kocaeli earthquake exceeds the TEC (2007) code limit of 0.20 m for the Minaret I model (Dogangun et al. 2008).

The most critical axial stress contours calculated during the dynamic analysis of minaret models are shown in Figure 15. The maximum tensile and compressive stresses occur at a height of 8 m, immediately above the transition zone (Figure 12). This is consistent with the most common minaret failure mode observed in the field after the earthquakes (Figures 7a, 8a and 8b). The minaret models used here were stand-alone fixed-ended cantilevers, which did have any lateral support over their height. Nevertheless, in some cases one side of the minaret boot or base (bottom 6 m of the minaret models) are supported by or attached to the mosque structure (Type II in Table 4, e.g., Figures 8a and8b). Such a partial lateral support, even on one side, provides added stiffness near the bottom of the minaret. This will further increase failure potential at the top of the transition zone as shown inFigures 7a, 8a, 8b and 14.

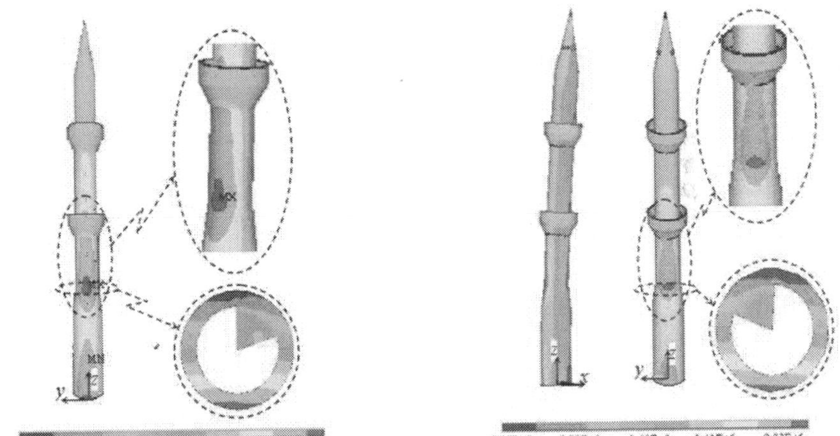

Figure 15. Maximum stress contours for Minaret-I under: a) Duzce, and b) Kocaeli records

SUMMARY AND CONCLUSIONS

Structural damage and failures observed in 59 historical and monumental structures (mosques and/or minarets) were documented after the August 17 and November 12, 1999 earthquakes. The reported damage distribution in minarets in the cities of Düzce and Bolu gives an indication of the extent of damage. Ten percent of all visited mosques collapsed while 20% suffered severe damage. Of all visited minarets, 38 percent collapsed (Figure 6). Five out of seven masonry minarets (approximately 70%) in the cities of Düzce and Bolu reporeted in this chapter collapsed (Table 4). Historical structures with heavy and stiff walls are subjected to larger lateral earthquake forces due to their small periods of vibration. The missing keystones in the upper-floor windows and wide cracks between the stones in the walls above and below window openings were the most common damage types.

The location of the failure in the minarets that collapsed during the 1999 earthquakes was generally at the region near the bottom of the cylinder where a transition was made from a square to a circular section. At the bottom of the cylinder body the lateral stiffness and strength are smaller compared with those of the transition region or the minaret base. Consistent with the observed damage, dynamic analysis results from three generic finite element minaret models the largest bending stresses in the same region of the minaret models. For minarets with the base or boot attached to the mosque structure (Type II), additional rigidity and stiffness provided by the mosque prevents deformation and damage within the base. In such cases the failure mostly occurs just above the transition region or base of the cylindrical body.

REFERENCES

1. Turkish Earthquake Code.1998and 2007).Specification for structures to be built in disaster areas; Part III-earthquake disaster prevention. Ministry of Public Works and Settlement, Government of Republic of Turkey.
2. Turkish Building Code. TS-5001985Building Code Requirements for Reinforced Concrete. Turkish Standards Institute, Ankara, Turkey (in Turkish).
3. H. Sezen, A. S. Whittaker, K. J. Elwood, K. M. Mosalam, 2003Performance of Reinforced Concrete and Wall Buildings during the August 17, 1999 Kocaeli, Turkey Earthquake, and Seismic Design and Construction Practice in Turkey. Engineering Structures, 25103114
4. Uniform Building Code.1997Structural Engineering Design Provisions, 2International Conference of Building Officials, ICBO. Whittier, California.
5. N. Çamlıbel, 1998Sinan Mimarlığında Yapı Strüktürünün Analitik İncelenmesi, Doçentlik Tezi, İTÜ, Fen Bilimleri Enstitüsü, İstanbul (in Turkish).

6. A. Ural, A. Dogangun, H. Sezen, S. Angin, 2012Seismic Performance of Masonry Buildings during the 2007 Bala, Turkey Earthquakes. Natural Hazards, 60310131026
7. H. Sezen, A. Doğangün, 2009Vulnerability and conservation of historical masonry structures. Prohitech 2009, Protection of Historical Buildings, Rome, Italy, June 2124
8. M. Oğuzmert, 2002Dynamic behaviour of masonry minarets, M.Sc. Thesis, Istanbul Technical University, Turkey.
9. A. Dogangun, H. Sezen, O. I. Tuluk, R. Livaoglu, R. Acar, 2007Traditional Turkish Monumental Structures and their Earthquake Response. International Journal of Architectural Heritage, 13251271
10. H. Sezen, G. Y. Firat, M. A. Sozen, 2003Investigation of the Performance of Monumental Structures during the 1999 Kocaeli and Duzce Earthquakes. Fifth National Conference on Earthquake Engineering, Istanbul, Turkey, 2630May
11. Y. G. Firat, 2001A Study of the Structural Response of Minarets in the 1999 Anatolian Earthquakes. M.S. Thesis. Purdue University. West Lafayette, Indiana.
12. A. Dogangun, H. Sezen, 2012Seismic Vulnerability and Preservation of Historical Masonry Monumental Structures. Earthquakes and Structures, 318395
13. H. Sezen, R. Acar, A. Dogangun, R. Livaoglu, 2008Dynamic Analysis and Seismic Performance of Reinforced Concrete Minarets." Engineering Structures, 30822532264
14. F. Portioli, O. Mammana, R. Landolfo, F. M. Mazzolani, L. Krstevska, L. Tashkov, K. Gramatikov, 2011Seismic Retrofitting of Mustafa Pasha Mosque in Skopje: Finite Element Analysis. Journal of Earthquake Engineering, 154620639
15. A. M. Turk, C. Cosgun, 2012Seismic Behaviour and Retrofit of Historic Masonry Minaret. Gradevinar, 641
16. A. C. Altunisik, 2012Comparison of earthquake behavior of reinforced concrete minarets using fiber-reinforced polymer composite. The Structural Design of Tall and Special Buildings, DOI:tal.731
17. C. S. Oliveira, E. Çaktı, D. Stengel, M. Branco, 2012Minaret behavior under earthquake loading: The case of historical Istanbul. Earthquake Engineering and Structural Dynamics, 411939
18. A. Dogangun, R. Acar, H. Sezen, R. Livaoglu, 2008Investigation of Dynamic Response of Masonry Minaret Structures. Bulletin of Earthquake Engineering, 63505517

CITATION

Halil Sezen and Adem Dogangun (2012). Seismic Performance of Historical and Monumental Structures, Earthquake Engineering, Prof. Halil Sezen (Ed.), ISBN: 978-953-51-0694-4, InTech, DOI: 10.5772/51338.

CHAPTER 3

Seismic Evaluation of Low Rise RC Framed Building Designed According to Venezuelan Codes

Juan Carlos Vielma1, Alex H. Barbat2, Ronald Ugel1 and Reyes Indira Herrera1

[1]Structural Engineering Department, School of Civil Engineering, Lisandro Alvarado University, Venezuela
[2]Department of Strength of Materials and Structural Analysis in Engineering, Technical University of Catalonia, Barcelona, Spain

INTRODUCTION

Along its history, Venezuela has been severely affected by destructive earthquakes [1]. Approximately 80% of the population lives in seismically active areas, where have occurred destructive earthquakes even in recent times [2]; The seismic hazard, inadequate design and construction of buildings as well as the damage occurred from previous earthquakes, demonstrate a high vulnerability in existing buildings. Then it is essential to continuously make progress and research in the field of earthquake engineering and upgrade the seismic design codes. Seismic upgrade requires the evaluation or predictions of the expected damage to structures at the time of an earthquake of a certain severity occur. From this prediction it can be defined solutions for the reduction of structural vulnerability [3].

The damage occurred in buildings after an earthquake indicates the need for reliable methodologies for the evaluation of seismic behavior of the existing buildings. According to current technical and scientific advances, seismic evaluation of reinforced concrete (RC) structures can be done by two different approaches: empirical methods and mechanical methods [4]. The current tendency of earthquake engineering in the evaluation of structural behavior is the application of simplified mechanical methods based on performance, involving the capacity spectrum [5], because there are developed refined models and detailed analysis.

This study used a mechanical method that involves non-linear analysis with deterministic and probabilistic approaches, as well as procedures of analysis based on Limits States defined by displacements [6], in order to evaluate the behavior of a low rise RC building with plan irregularity, designed according to Venezuelan codes [7]-[9] and subjected to seismic action effect. Through the use of mathematical models and computational tools, seismic behavior of the building is obtained in a suitable way. Among these tools any procedure was chosen: the quadrants method, which leads to the rapid assessment of the seismic capacity of a structure through its non-linear response [10]. Results of the research shown that the current design of this kind of structures is not safe when they are under the maximum seismic actions prescribed by codes, then it is necessary to review the design procedures in order to find more realistic designs that fulfill the goals of the performance-based design.

CASE STUDIED

A two story RC framed building was studied, (Figure 1a), which contains internal staircase and 220 m^{2}total plan area. This structure represents a common typology used for residential buildings in Venezuela, for prone seismic zones. The structure was designed and detailed for a high ductility value (response reduction factor of 6).

The building was modeled according its original design, called *original building (OB)*, with plan asymmetry (Figure 1b) and one way 25 cm depth slabs in *X* direction. A second model was designed adjusted to seismic performance requirements formulated by Herrera *et al.* [12], called *resizing building (RB)*, which presents equal geometrics and mechanics characteristics than *OB* model, but considering the "strong column-weak beam" condition. It was also used the displacement-based seismic design procedure of *Priestley et al.* [13] in order to design of a third model, called*displacement-based design building (DBDB)*. These three models differ only in the dimensions of its structural elements (Table 1).

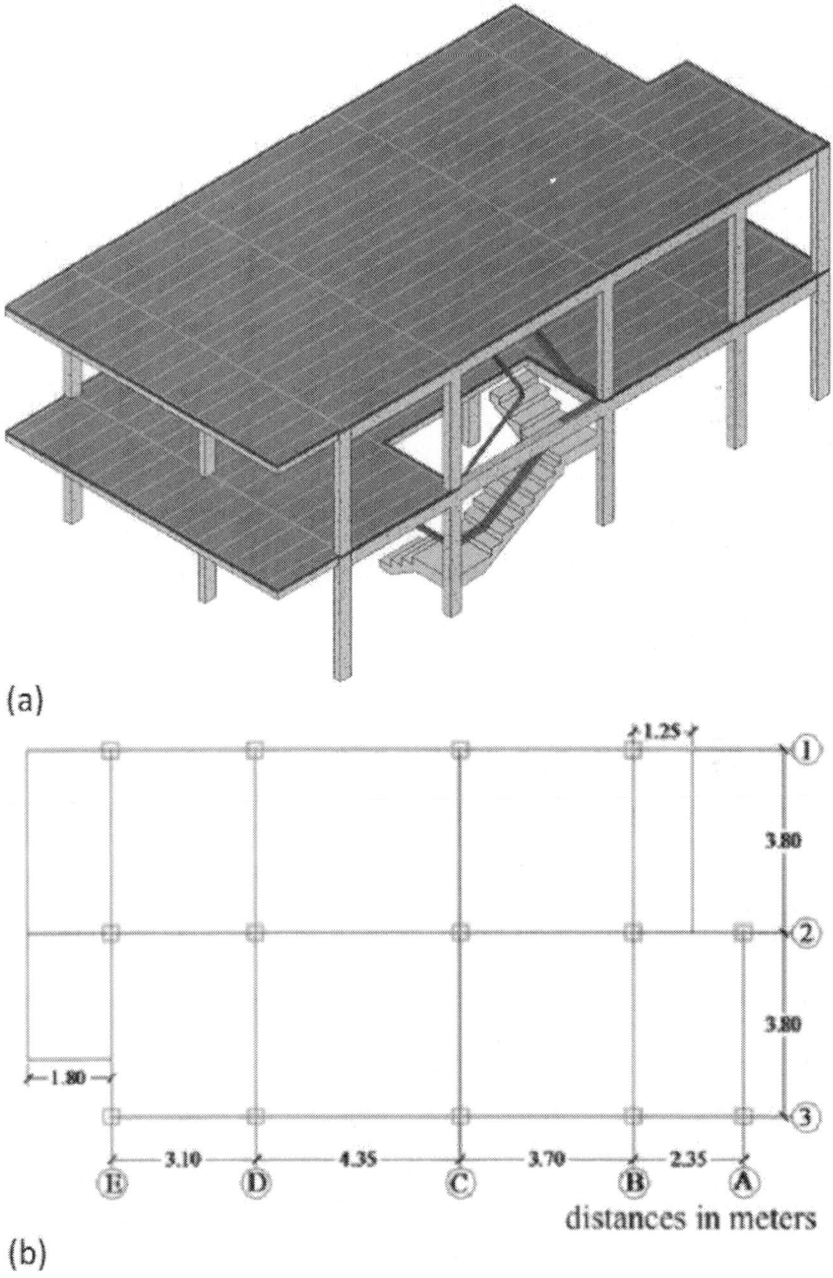

Figure 1. Low rise RC building(a) 3D view (b) Plan view

Table 1. Geometric characteristics of elements from each modeled building

Building	Axis X beams (cm)	Axis Z beams (cm)	First level columns (cm)	Second level columns (cm)
OB	20x35	20x35	20x30	20x30
RB	20x45	20x35	30x30	30x30
DBDB	20x40	20x40	35x35	30x30

ASSESSMENT METHOD

The Quadrants Method is based on the results of the non-linear static analysis (Pushover analysis). This analysis results are plotted in a displacement vs. base shear format, this generate the capacity curve which represents the overall capacity of the whole structure against lateral forces. In order to evaluate the capacity curve two of the main structural parameters are taken into account. The first one is the design elastic shear, obtained from the elastic analysis of the structure using the elastic design spectrum. The second parameter is the threshold that defines the Repairable Limit State, obtained from [14] for RC framed buildings with similar characteristics to the studied ones. The thresholds have been computed from characteristic values of three levels of damage proposed in [15] and are showed inTable 2. Both values are used to define two axes over the capacity curve, the elastic base shear defines an horizontal axis and the damage threshold defines a vertical axis, then Capacity Curve is divided in four spaces or quadrants, see Figure 2.

Table 2. Inter-storey drifts adopted for the damage thresholds determination

Limit State	Damage type	Seismic hazard	Probability event	Inter-storey drift in (%)
Service	Non-structural	Frequent	50% in 50 years	$0,2<\delta<0,5$
Damage control	Moderate structural	Occasional	10% in 50 years	$0,5<\delta<1,5$
Collapse prevention	Severe structural	Rare	2% in 50 years	$1,5<\delta<3,0$

The performance point is a common procedure accepted among the scientific community to evaluate the seismic performance of a structure under a specific demand. It is usually obtained from the idealized shape of the Capacity Curve as is shown in the Figure 2 [16]. The Quadrants Method also uses this parameter in order to define the roof displacement of the case studied, defined according to the N2 method [5]. If the performance point is under the axis defined by the elastic base shear (Quadrants III or IV), the design does not meet the basic objective of the seismic design because the building does not have enough lateral strength. If the performance point is on the right side of the vertical axis (Quadrant I) means that the building has adequate stiffness, otherwise (Quadrant II) it means that the stiffness is very low and the displacements can be longer than the displacements that can produce advanced structural damage, technically or economically irreparable. These lateral displacements are usually computed from the dynamic response of the structure submitted to a strong motion with a return period of 475 years, or an occurrence probability of 10% in 50 years [17].

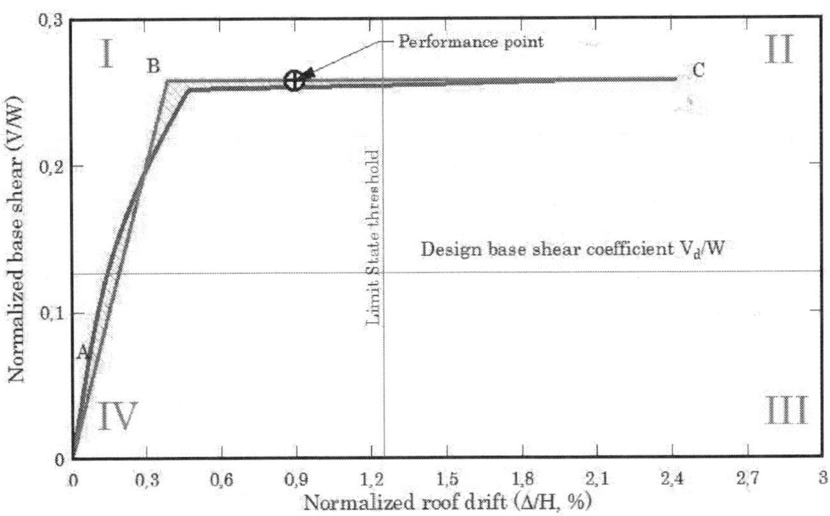

Figure 2. Capacity curve and the axis that define the Quadrants Method

The Quadrants Method can provide an objective criterion in order to upgrade the seismic capacity of a structure. If the performance point is on the Quadrant I, the structure has enough lateral strength and stiffness, so does not need to be reinforced. If the structure is on the Quadrant II, it is necessary to provide additional stiffness by using conventional procedures like RC or steel jacketing. If the performance point is on Quadrant III, the structure requires a more radical intervention, adding stiffness and lateral strength. In this case it is possible to combine some traditional reinforcement techniques with new ones like FRP jacketing. In this case the columns are the subject of the main intervention. Finally, if the performance point is on the Quadrant IV, the structure does not has enough lateral strength and then the reinforcement technique must be FRP jacketing.

NONLINEAR ANALYSIS

The structures are modeled by incorporating the structural response when it incurs in the material and geometrical non-linear range, produced by high deformations caused by accidental excitations (earthquakes) [11]. The analyses were performed using ZEUS-NL software [18], which allows to model complex structures with "n" number of finite elements, thus to know the elements in the building which are most vulnerable to damage. Each building is modeled in two dimensions, spitting each frame to get a more detailed response for the seismic behavior of each frame; a 3D dynamic analysis was applied to the ER model.

The static Pushover analysis is performed once the frames have been subjected to action of gravity loads, based on the pseudo-static application of lateral forces equivalent to displacements of seismic action [5]. The pattern of lateral seismic loads consist in increasing loads with height (triangular distribution) applied in a monotonic way until the structure reaches its maximum capacity [20].

This procedure applies a solution of equilibrium equations in an incremental iterative process form. In small increments of linear loads, equilibrium is expressed as:

$$K_t \Delta_x + R_t = \Delta F \tag{1}$$

Where *Kt* is the tangent stiffness matrix, *Rt* is the restorative forces at the beginning of the increased load. These restorative forces are calculated from:While this procedure is applied, the strength of the structure is evaluated from it is balance internal conditions, updating at each step the tangent stiffness matrix. Unbalanced loads are applied again until it can satisfy a convergence criterion. Then, a new load increase is applied. The increases are applied until a predetermined displacement is reached or until the solution diverges.

From the capacity curve provided in this analysis, it is determined the structural ductility (μ) by the quotient between thc ultimate displacement and cadence point displacement, as shown in the following expression:

$$R_t = \Sigma K_t, K\Delta_u \tag{2}$$

Where Δ_u is Ultimate displacement and Δ_y is the global yield displacement. Both values are computed from the idealized capacity curve of the structure.

By the other hand, the dynamic analysis is an analysis method that can be used to estimate structural capacity under seismic loads. It provides continuous response of the structural system from elastic range until it reaches collapse. In this method the structure is subjected to one or more seismic records scaled to intensity levels that increase progressively. The maximum values of response are plotted against the intensity of seismic signal [21-22]. The procedure to perform the dynamic analysis from the seismic signal is:

- To define a seismic signal compatible with the design scenario;
- To define the scaled earthquake intensity a monotonic way;
- To define the extent of damage or damage Limit States;
- To study a seismic record for the dynamic analysis of a structural model parameterized to measure earthquake intensity;

The non-linear dynamic analyses provide a set of curves which are a graphical representation of the evolution of the drifts respect time. Results let to compute the damage lumped in specific elements of the structure, but these results are beyond the objective of this Chapter.

Table 3. Limit States and seismic hazard level

Analysis earthquake	**Limit State**	**Return period (years)**	**Occurrence probability in 50 years**	***Interstorey drift δ (%)***
Frequent	Serviceability	95	50%	δ < 0,5
Rare	Reparable damage	475	10%	δ < 1,5
Very Rare	Collapse Prevention	2475	2%	δ < 3,0

For the dynamic analysis the structures were subjected to seismic action (see Table 2) defined by accelerograms built on the basis of a likely value of maximum acceleration of the soil and the hazard level associated with the location of the structure and other seismic characteristic design parameters [16]. These accelerograms called "synthetic accelerograms" are generated through the implementation of a set of earthquakes with wide frequency content, using the PACED program [17], based on the Venezuelan code's elastic design spectrum. For the dynamic analyses of the three buildings (*OB, RB, DBDB*), it were used 3 synthetic accelerograms with duration of 60sec.

Non-linear dynamic analysis was applied to all buildings in order to verify if the performance evaluated by the Quadrants Method is reliable in order to evaluate the fulfilment of the thresholds defined in the precedent section. For this purpose they has been computed three synthetic elastic design spectrum-compatible accelerograms by means of the PACED program [23]. In Figure 3 are shown the Venezuelan rigid-soil elastic design spectrum with the response spectra obtained from the synthetic accelerograms.

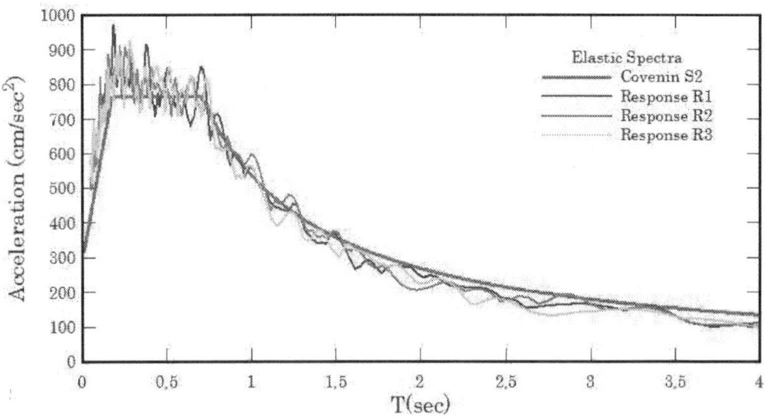

Figure 3. Elastic response spectra from elastic design spectrum-compatible accelerograms

These three earthquakes were applied to all frames from the three buildings evaluated, in order to obtain maximum displacement that can be reached by each one. In the software used [18], it was required the implementation of dynamic loads in direction X and the assignation of a control node located in the gravity center of the roof level.

The 3D non-linear dynamic analysis is based on the procedure explained in [20]. The RB building is analyzed, defining its geometry, materials and sections, serviceability loads in Y direction in all beams-columns joints, and dynamic loads on outer nodes with directions and combinations shown inTable 4. One direction ribbed slabs were modeled as rigid diaphragms in its plane by using additional elements with no flexural capacity (Figure 4).

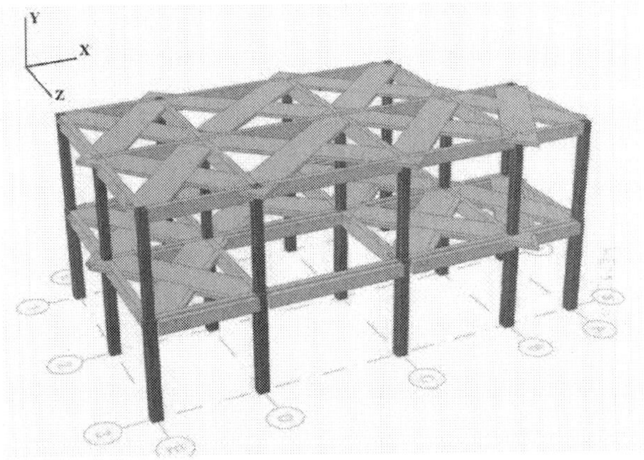

Figure 4. Rigid diaphragms in 3D *RB* framed building

Once built the model, there were applied all the accelerograms with the combinations shown in Table 3, for the interstorey drifts and maximum torsional moments on supports. These combinations are based on the Venezuelan seismic code [7] and following established by [24] about the seismic response of asymmetric structural systems in the inelastic range.

Table 4. Applied seismic combinations.

Nº	Seismic combination
1	100 % (X)
2	100 % (Z)
3	100% (X) y 30% (Z)
4	100% (Z) y 30% (X)

RESULTS

From classic elastic analysis, the verification of interstorey drifts of the *OB* building, shows that they exceed the limit established in [7], while in the *RB* model it was obtained that it meets the code's parameters, which limits the inter storey drift to 0,018. By the other hand, in the *DBDB* building were not performed drifts verifications, since it was designed based on the method performed in [13], where the generated seismic forces are originally limited to not exceed the limit value of drift specified in the applied code.

To determine the values of structural ductility it was necessary to plot the idealized curve in function of the capacity curve obtained from non-linear pseudo-static analysis (pushover analysis), in order to know the point at which the structure begins to yield. Figure 5 shows an example of the normalized capacity curve with the idealized (bi-linear) curve of Frame C of *OB*. Structural ductility for each evaluated building values are presented in Table 5. From this Table it is evident that the original building, designed according to current Venezuelan codes has ductility values lower than the redesigned and the displacement-based buildings.

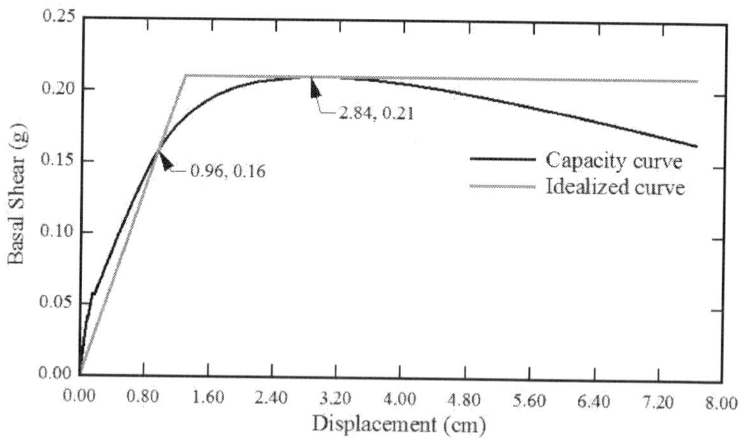

Figure 5. Normalized and idealized capacity curves. Frame C. *OB* building

Table 5. Structural ductility results

	Frame	Building EO	ER	DBDB
Structural Ductility	**A**	5,56	5,52	4,77
	B	2,22	6,04	5,38
	C	2,17	4,69	5,25
	D	2,21	5,54	5,59
	E	2,23	7,07	6,06
	1	2,66	5,29	6,69
	2	2,20	4,17	5,92
	3	2,83	5,95	6,24

From the obtained capacity curves there were computed the Performance point (Pp) of every frame of each evaluated building. Table 5 presents the values of Pp of all the frames of evaluated buildings.Figure 6 shows the determination of the Pp of Frame C from *OB* building using the N2 procedure proposed by Fajfar [5].

Table 6. Performance points (Pp) of studied buildings frames

	Pp (cm)		
FRAME	EO	ER	*DBDB*
A	5,94	2,42	2,52
B	13,89	9,47	7,43
C	15,22	9,50	9,38
D	14,01	9,50	7,57
E	13,45	9,55	6,60
1	12,62	9,35	6,07
2	15,74	11,48	9,29
3	10,92	7,57	4,23

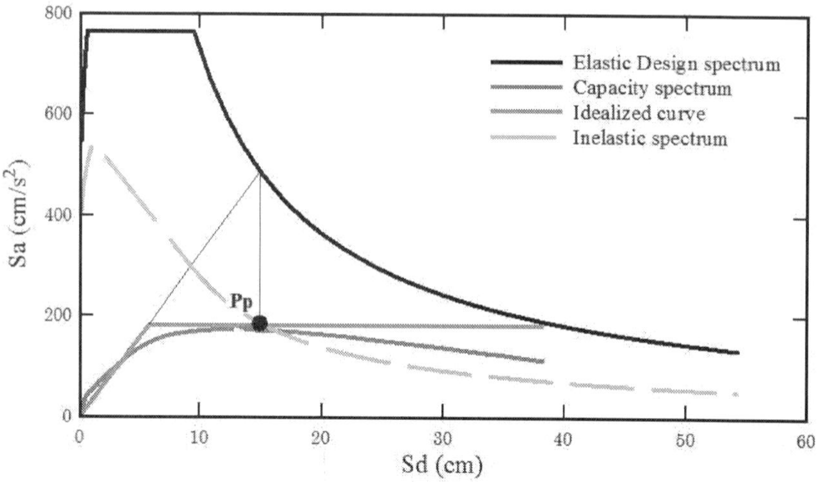

Figure 6. Performance point of Frame C. *OB* Building, determined by N2 procedure

From dynamic analyses, there were determined global and inter s to re y drifts of each frame from all three models studied. Both types of drifts were calculated on the basis of the application of synthetic ace le r o grams with different intensities, representing the lateral forces applied to frames in order to generate their respective maximum displacements. Figures 7 to 9 show the evolution of the global (Δ/H) drifts expressed as a percentage, respect to time (sec) of the frame C from *OB*, *RB* and *DBDB* models for a peak ground acceleration of 0,3 g, respectively.

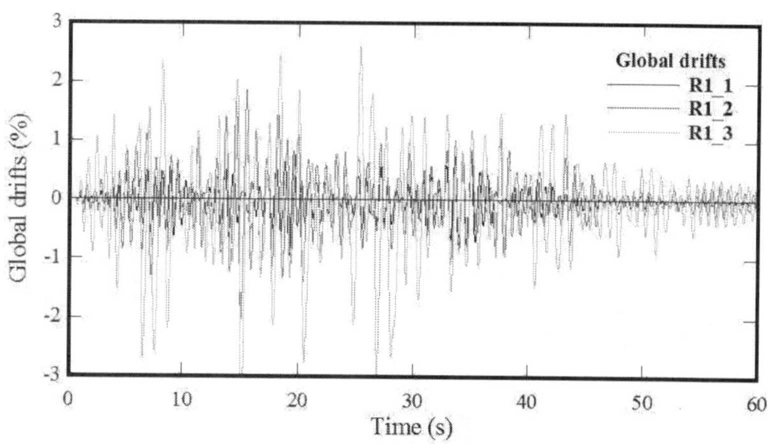

Figure 7. Global drifts. R1 earthquake. Frame C. *OB* building.

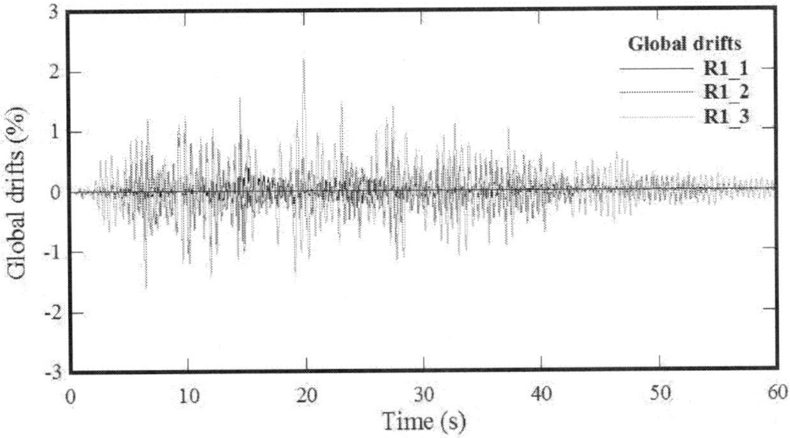

Figure 8. Global drifts. R1 earthquake. Frame C. *RB* building

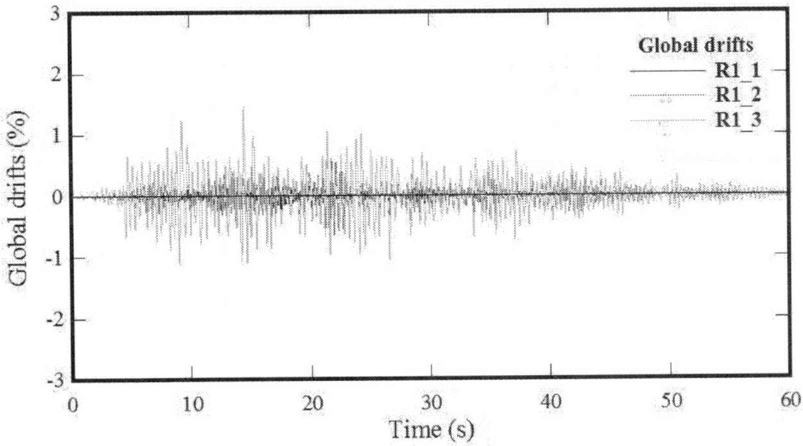

Figure 9. Global drifts. R1 earthquake. Frame C. *DBDB* building

Figures 10 to 12 show the results for interstorey drifts of frame C from *OB*, *RB* and *DBDB* buildings, taking into account the R1_3 earthquake with duration of 60 seconds. Similarly, interstorey drifts for applied earthquakes, R1, R2 and R3 with its three intensities, were obtained. It were verified for each Limit State considered in this study. Table 6 reflects the values of interstorey drifts of buildings in study for earthquake R1, taking into account the three levels of hazard, 0,5%, 1,5% and 3%, for the Limits States considered.

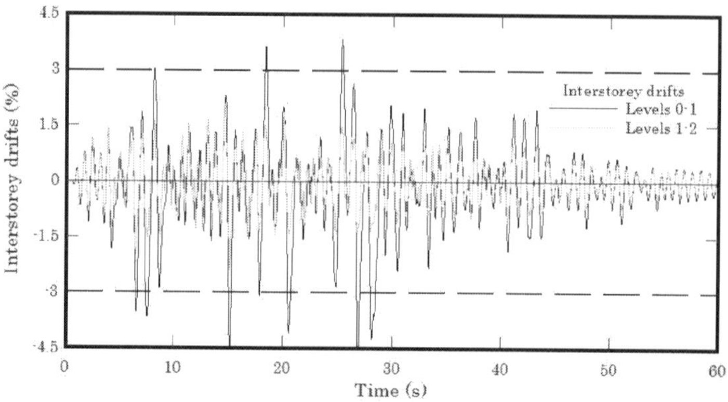

Figure 10. Interstorey drifts. R1_3 earthquake. Frame C. *OB* building

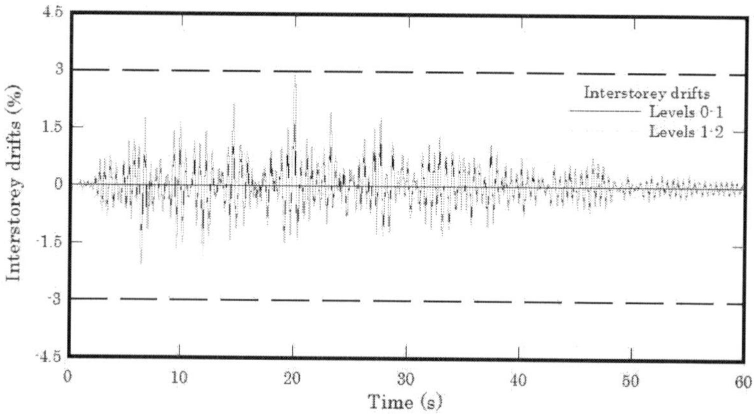

Figure 11. Interstorey drifts. R1_3 earthquake. Frame C. *RB* building

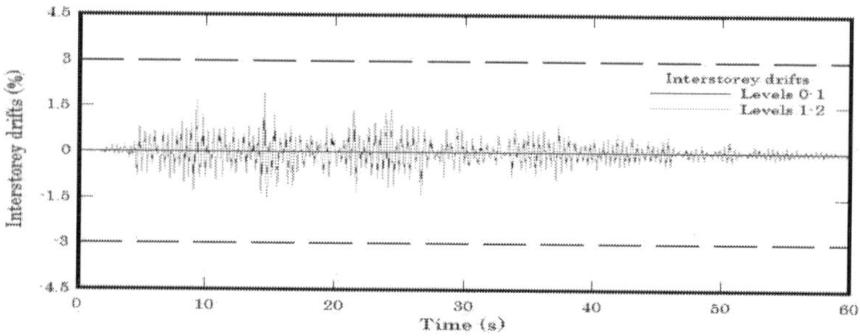

Figure 12. Interstorey drifts. R1_3 earthquake. Frame C. *DBDB* building

Table 7. Interstorey drifts verification. R earthquake. *OB*, *RB* and *DBDB* building

	LIMITS STATES								
	OB			**RB**			***DBDB***		
FRAME	**SLS**	**RDLS**	**PCLS**	**SLS**	**RDLS**	**PCLS**	**SLS**	**RDLS**	**PCLS**
A	-	-	-	-	OK	OK	OK	OK	OK
B	-	-	-	OK	OK	OK	OK	OK	OK
C	-	-	-	OK	OK	OK	OK	OK	OK
D	-	-	-	-	OK	OK	OK	OK	OK
E	-	-	-	-	OK	OK	OK	OK	OK
1	-	-	-	-	OK	OK	OK	OK	OK
2	-	-	-	-	OK	OK	OK	OK	OK
3	-	-	-	-	OK	OK	OK	OK	OK

[i] - **SLS**: Serviceability Limit State, **RDLS** Reparable damage Limit State, **PCLS**: Prevention of Collapse Limit State; -: No meet the norm,+: Checks the Venezuelan seismic code

3D Nonlinear dynamic analysis

Interstorey drifts in frames of *RB*, were obtained by applying the R1_3 earthquake for the combinations 1 and 2 (Table 7). According to results obtained, interstorey drifts in 2D and 3D modeled buildings differ greatly from each other, for this reason it is important to take into account the 3D analysis in order to evaluate the drifts of buildings, because irregularities can produce lateral displacements that does not match with the obtained in 2D analysis.

In the 2D model greater drifts were obtained, while in 3D model the drifts were reduced by the contribution of the diaphragms. Also it were determined the maximum torsional moments in each column before the implementation of R1_3 earthquake in all supports for the four combinations described in Table 8. In Figure 13 have been plotting torsional moments in function of time for the four combinations, where nodes appointed by n111 until the n513 are corresponding to supports, while Figure 14 shows the maximum torsional moment range for each column from three-dimensional analysis. It is evident that for the accelerograms used, the maximum torsional moments occurs in the extreme columns and in the columns located in the intersection of the structure. This is an important feature that confirms the negative effect of the irregularity combined with the seismic action. The torsional moments for the other seismic combinations used in this study were obtained using the same procedure.

Table 8. Maximum torsional moments for seismic combinations.

Seismic combination	Node-column Description	Max. Torsional Moment. (Nxm)
1	Corner column. n513	64225
2	Corner column. n512	76000
3	Corner column. n513	41000
4	Corner column. n512	65000

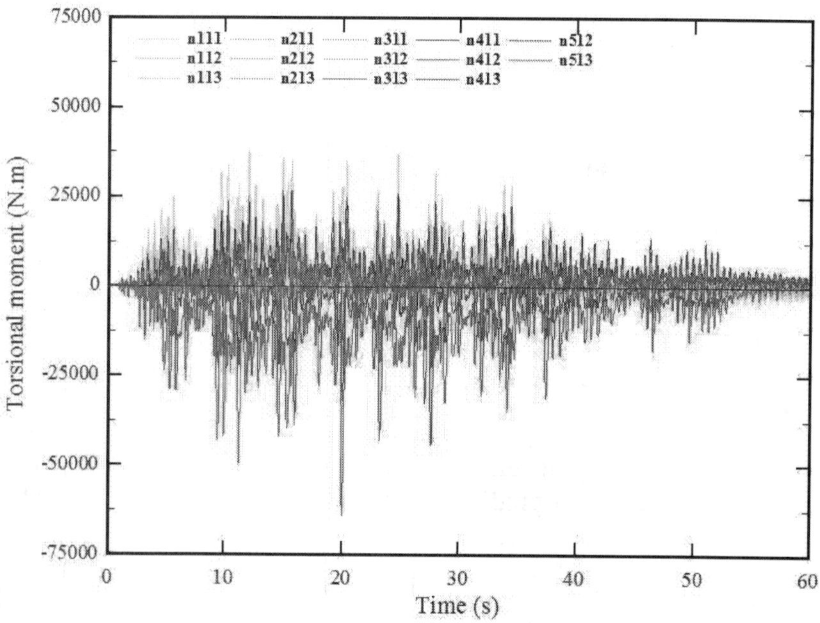

Figure 13. Torsional moments for earthquake 100% (X.)

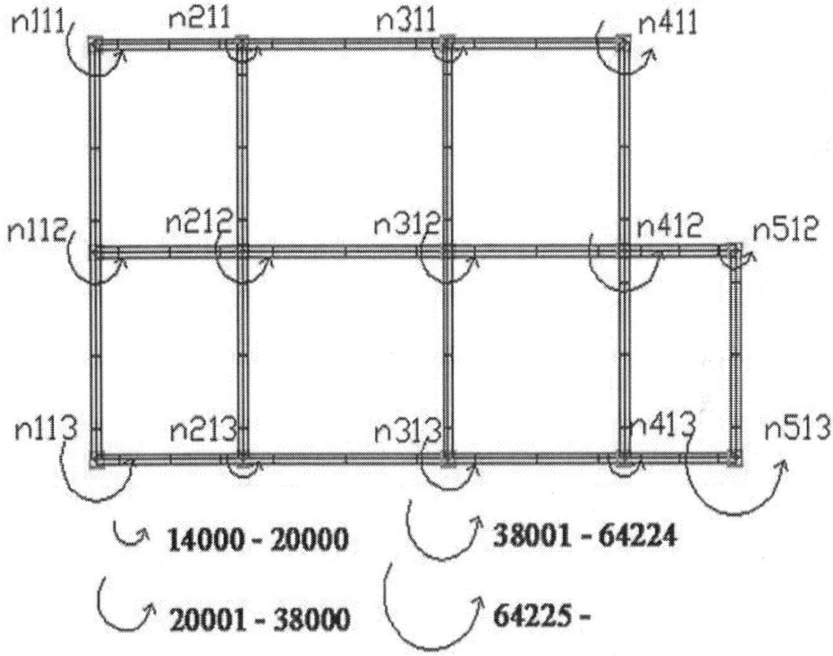

Figure 14. Torsional moments for earthquake 100% (X). Plant detail.

CONCLUSIONS

In order to know the seismic response of the studied building it were used analytical methods considering the seismic hazard level and structural regularity criteria. The elastic analysis applied to the *OB* building identified elastic displacements greater than maximum value of interstorey allowed by Venezuelan seismic code. From the resizing model *RB* it was obtained interstorey drifts that satisfied the maximum value established in the code. Thus, the sections of the structural elements of *OB* are insufficient to properly control the damage caused by seismic forces.

From dynamic analysis there were computed the global and interstorey drifts for all three evaluated models determining the dynamic response of these structures and controlling the damage level reached in them. With the global drifts, it was evaluated the threshold of the collapse Limit State, which corresponds to the maximum value of 2.5%. *RB* and *DBDB* buildings reached drifts values below this limit, proving good seismic performance on both buildings; *OB* presented drifts values which exceeded this limit. In the verification of interstorey drifts it was generally noted that interstorey drifts of *OB* building were longer than the considered by hazard levels, while the two resized buildings reached values within the thresholds established for each Limit State.

Three-dimensional dynamic analysis applied to *RB* building allowed determine that interstorey drifts values were under the threshold of the Limit States considered. On the other hand, in order to know the maximum torsional moments for each column in this model, there were applied four seismic combinations where it was noted that there was greater torsion in the case of the component of the earthquake in Z-direction. Based on these results it was demonstrated the structural asymmetry of the assessed building since the center of mass does not coincide with the center of rigidity, determining that the greatest torsional moments are on outer columns and inner corners.

Interstorey drifts of *RB* building obtained from 2D and 3D nonlinear dynamic analysis, it was noted that 2D model provided greater drifts values than the 3D model drifts. This is a logical and expected result since the 3D dynamic analysis considers the rigid diaphragm, which introduces restrictions to the number of degrees of freedom in the structure.

Inelastic static analysis is more reliable than linear methods in the prediction of the parameters of response of buildings, although this method has no response on the effects of higher modes of vibration. A more reliable and sophisticated method is the 2D no linear dynamic analysis, where it can be better determined the likely behavior of the building in response to the earthquake. However, the uncertainties associated with the definition of accelerograms used in these analysis and properties of coplanar structural models can be reduced with the implementation of the dynamic 3D analysis because there are considered factors associated with structural redundancy and are used more actual values in terms of rigidity of resistant structural lines.

The Quadrants Method presented in this paper is suitable for rapid and reliable evaluation of structures, with a low calculation effort. The cases studied demonstrated that the method can provide a reliable criterion to predict if any structure would have an inadequate seismic performance based on the results of static non-linear analysis. Results obtained from dynamic non-linear analysis confirmed the results obtained from the application of the Quadrants Method.

Despite the plan irregularity of the studied building, the Quadrants Method was suitably in order to predict that its lateral stiffness was not enough. Dynamics analysis confirmed this feature, then the cross sections of this building was resizing and details of the confinement were improved in order to meet the regulations of the current version of the Venezuelan seismic code. The seismic performance of the new designed building was tested with the Quadrants Method and dynamic analysis, showing that the resizing structure met all the Limit States used in this research.

NaN. ACKNOWLEDGEMENTS

Authors wish to acknowledge to the Scientific and humanistic Council of Lisandro Alvarado University for the financial support of the research in the field of non-regular structures. Also the authors would highlight the role of the International Center of Numerical Methods for Engineering in the collaborative research.

REFERENCES

1. Grases J., Altez R. and Lugo M. Destructives Earthquakes Catalogue. Venezuela 1530/1998. Central University of Venezuela. Natural Sciiences, Physics and Mathematics Academy. Engineering School. Caracas, Venezuela, 1999.
2. Pérez O. and Mendoza J. Seismicity and tectonics in Venezuela and surroundings areas. Earth Physics,1998, 10, pp 87-110.
3. Bar bat A., Mena U. and Yépez F. Probabilistic evaluation of seismic risk in urban zones. International magazine for numerical methods for Calculus and engineering projects. 1998, 14, 2, 247-268.
4. Calvi, G., Pinho, R., Magenes, G., Bommer, J., Restrepo, L and, Crowley, H. Development of Seismic Vulnerability Assessment Methodologies over the Past 30 Years, ISET Journal of Earthquake Technology, 2006, Paper No. 472, 43, 3, 75-104
5. Fajfar P. Nonlinear analysis method for performance based seismic design. Earthquake Spectra, EERI, United States of America, 2000, 16, 3, 573-591.
6. Vielma, J. C., Barbat, A. H. and Oller, S. Framed structures earthquake resistant design. International Center for Numerical Methods in Engineering (CIMNE) Monograph. Earthquake Engineering Mongraphs. Barcelona, Spain, 2011.
7. Covenin 1756:01. Earthquake-resistant Design code. Part 1. Fondo Norma. Caracas, Venezuela, 2001.
8. Covenin 1753:06. Design and construction of buildings with structural concrete. Fondo Norma. Caracas, Venezuela, 2006.
9. Covenin 2002:88. Minimum design loads and criteria for buildings code. Fondo Norma. Caracas, Venezuela, 1988.

10. Vielma, J. C., Barbat, A. and Martínez, Y. The Quadrants Method: A procedure to evaluate the seismic performance of existing buildings. 15 World Conference on Earthquake Engineering. Lisbon, Portugal, 2012.
11. Herrera, R., Vielma, J. C., Ugel, R., and Barbat, A. Optimal design and earthquake resistant design evaluation of low rise framed rc building. Natural Science, Earthquake Special Issue. August 2012. California, USA. Doi:10.4236/ns.
12. Vielma, J. C., Barbat, A. H. and Oller, S. Seismic safety of RC framed buildings designed according modern codes. Journal of Civil Engineering and Architecture. 2011, 5:7, 567-575. David Publishing Company. Chicago, United States of America.
13. Priestley M., Calvi G. and Kowalski M. Displacement-based seismic design of structures. IUSS Press. Pavia. Italia, 2007.
14. Vielma, J. C., Barbat, A. H. and Oller, S. Nonlinear structural analysis. Application to evaluating the seismic safety. Nova Science Publishers. New York,2009.
15. Di Sarno, L. and Elnashai, A. Fundamentals of Earthquake Enginering. John Wiley and Sons. Chichester, United Kingdom, 2008.
16. Park, R. State-of-the-art report: ductility evaluation from laboratory and analytical testing. Proceedings 9th WCEE, IAEE, Tokyo-Kyoto, Japan, Vol VIII: 605-616, 1988.
17. Vielma, J. C., Barbat, A. H. and Oller, S. Seismic sizing of RC buildings according to energy-based amplification factors (In Spanish). Hormigón y acero. 2011. 63:263, 83-96.
18. Elnashai A., Papanikolau V. and Lee, D. ZEUS-NL, A system for Inelastic Analysis of Structures. User Manual. Mid-America Earthquake Center report no. MAE, Illinois University. Urban, Champagne, Illinois, 2011.
19. Papanikolaou, V. and Elnashai, A. Evaluation of conventional and adaptive pushover analysis I: methodology, Journal of Earthquake Engineering, 2005, 9, 6, 923-941
20. Mwafy, A. and Elnashai, A. Static pushover versus dynamic collapse analysis of RC buildings, Engineering Structures, 2001, 23, 407–424
21. Vamvatsikos D. and C. Allin Cornell (2002). Incremental dynamic analysis, Earthquake Engineering and Structural Dynamics, 2002, 31, 491–514.
22. Kappos, A. and Stefanidou, S. A deformation-based seismic design method for 3D R/C irregular buildings using inelastic dynamic analysis. Bulletin of Earthquake Engineering, Springer, Netherlands, 2010, 8, 4, 875-895
23. UCLA-CIMNE. Compatible accelerograms with elastic design spectrums generation software. (PACED). International Center for Numerical Methods in Engineering. Universidad Centroccidental Lisandro Alvarado. Venezuela, 2009.

24. Fajfar, P., Marusic, and D., Perus, I. Torsional effects in the pushover-based seismic analysis of buildings. Journal of Earthquake Engineering, 2005, 9, 6, 831-854.

CITATION

Juan Carlos Vielma, Alex H. Barbat, Ronald Ugel and Reyes Indira Herrera (2013). Seismic Evaluation of Low Rise RC Framed Building Designed According to Venezuelan Codes, Engineering Seismology, Geotechnical and Structural Earthquake Engineering, Dr. Sebastiano D'Amico (Ed.), ISBN: 978-953-51-1038-5, In Tech, DOI: 10.5772/55158

CHAPTER 4

Field Investigations of High Stress Soft Surrounding Rocks and Deformation Control

Weijian Yu1, 2, Weijun Wang1, Xinyuan Chen2, Shaohua Du2

[1]Hunan Provincial Key Laboratory of Safe Mining Techniques of Coal Mines, Hunan University of Science and Technology, Xiangtan, Hunan 411201, China
[2]School of Energy and Safety Engineering, Hunan University of Science and Technology, Xiangtan, Hunan 411201, China

ABSTRACT

Field investigations of high stress soft rock deformations show that the high stress soft rock roadway can slide with large deformation. Severe extrusion and floor heave can also be subsequently observed. The supported roadway can be locally damaged or completely fail, where the floor has a large deformation and/or is seriously damaged. The factors inducing large deformation of surrounding rocks in deep roadway are rock strengths, structure face cutting types, stress states, stress release, support patterns, and construction methods. Based on the deformation characteristics of high stress soft rock roadway, a comprehensive support scheme is proposed. The overall support technology of "step-by-step and joint, hierarchical reinforcement" for roadway is presented, and the anchor cable and bolt parameters to check the design methods are also given. Finally, the proposed comprehensive support method "bolt + metal mesh + U-steel arch + shortcrete + grouting and cable" is used in the extension section of east main haulage roadway at −850 m level of Qujiang coal mine. The 173-day monitoring results show that the average convergence of sidewalls reaches 208 mm, and the average relative convergence of roof and floor reaches 448 mm, suggesting that this kind of support technology for controlling large deformation of high stress soft surrounding rock roadway is effective.

INTRODUCTION

Under high in-situ stress, the soft surrounding rocks are basically characterized by great deformation rate, large and long-term deformation. At present, the deformation control technology of soft surrounding rocks in high in-situ stress environments is a hot and challenging issue. With respect to deformation mechanism, Chen et al. (2011) used the Lagrange multiplier method to model large deformation and nonlinear frictional contact of excavation boundary in a deep soft rock tunnel. The deformation of the transportation tunnel was asymmetrical due to the inclination of rock strata. The procedures for design of supports and stabilization measures in roadways were described by Šňupárek and Konečný (2010). He et al. (2007) summarized the causes of supporting failure of deep soft rock roadway as follows: low strength of wall rock, inadequate support, water effect, and structure stress effect. Kanji (2014) discussed the efforts made to investigate soft rocks, and stated that soft rocks have intermediate strength between soils and hard rocks. Gonzalez-Nicieza et al. (2008) adopted a numerical method to analyze convergence and constraint procedure for different section shapes and overburden depths. Hayano et al. (1999) suggested that the most typical characteristic of deep soft rocks is the long-term creep behavior, and discussed it through the observations of laboratory experiment. Bai et al. (2007) used the exponential creep model to analyze the changing behavior of stress and deformation rate of surrounding rocks with elapsed time, the effects of secondary supporting time and the intensity of surrounding rocks on deformation. With respect to the control of concentrated stress in surrounding rocks and reduction of rock crushing area, Wang et al. (2011) argued that the main reason of severe deformation is due to the active pressure and low intensity of the original support. Verma and Singh (2010) discussed the stability of tunnel and stress pattern, and performed numerical analyses for stress–strain distribution of tunnel excavation and associated support systems.

As for the control measures, Wang et al. (2011) proposed a supporting technology of flexible layer with rigid gap (He et al., 2005), the large deformation coupling support (He et al., 2013) and the anchor bolt with constant resistance and large deformation. The comprehensive support methodof"bolt + metalmeshes + pouring cover + anchor + grouting + floor anchor" was selected to rework the main roadway (Yu et al., 2014), based on the overlap arch bearing body strength theory for deep soft surrounding rocks (Yu et al., 2010). In addition, large deformation control of soft surrounding rocks of roadway with high stress and stability of support structure are important issues associated with increasing support strength (Xie et al., 2010).

The cases studied are mainly concerning deep rock mass deformation characteristics, for which reasonable control measures are proposed. However, due to complex geological conditions and lithology differences, the deformation characteristics of surrounding rocks in deep mining are not fully understood, which sometimes makes the design of control measures and support parameters impractical, resulting in serious deformation and/or failure. In this regard, we examine several typical haulage ways of deep mines in Jinchuan, Qujiang and Pingdingshan with respect to failure modes based on site investigations. The investigations include rock strength, structure face cutting type, stress state, support pattern, construction method, and other factors that have impacts on deep roadway stability to provide real data for roadway support design and control measures adopted.

ROADWAY FAILURE MODES

The details of typical deep mines in Jinchuan, Qujiang, and Pingdingshan are shown in Table 1.

Table 1. Basic information of investigated mines.

Mine	Description of geographic location	Mining depth (m)	Description of main surrounding rock conditions	Rock strength (MPa)	In-situ stress (MPa)	
					The vertical stress	The maximum horizontal stress
Jinchuan mine in Gansu Province	It is located in west of Jinchang City, middle east part of Hexi Corridor, i.e. in the southwest edge of Tengger Desert, north of Longshou Mountain in northwest direction. The topography is middle mountain. The altitude is 1600–1800 m	800–1000	The surrounding rock is homogeneous mixing dike, mixed thin gray rock of migmatite and amphibolite. The shale from rock crushed structure is significantly developed. The rest ore was affected by the metamorphism of fault zone. The mineral components become very complex and have mudding phenomena after excavation	13.5–25.5	20.3–26.8	35.9–50.2
Qujiang coal mine in Jiangxi Province	It is located in the northeast of Fengcheng City, low and flat, i.e. the Ganjiang River from southwest to northeast across the South Mine. The altitude is 22–64 m	850–900	The surrounding rock lithology is siltstone, black, thin to medium thick bedded, mixed with thin mudstone and fine sandstone belt. It contains a small amount of siderite concretions and plants fossil. The joint is developed with calcium-shale cementation	19.3–45.5	26.1–30.6	31.3–42.8
Pingdingshan mine in Henan Province	It is located in the west of Henan Province, south of Ruhe River, north of the Shahe River—Funiu Mountains region of low hill zone, characterized by high terrain in the northwest and low in the southeast. The altitude is 85–377 m	800–900	The surrounding rock is thick mudstone and sandy mudstone, occasionally sandstone or pebbly sandstone. Shallow joint is developed with space of 0.5–1.0 m, and joints in depth are sparsely distributed. Rock composition contains clay minerals, such as kaolinite montmorillonite, and softening and mudding can easily be induced by water	12.5–31.9	22.4–27.3	34.2–49.3

Soft rocks under high in-situ stress can lead to overall instability of surrounding rocks. The soft rock deformation characteristics under high in-situ stress can be summarized as follows:

1) The two sidewalls of tunnel were severely squeezed under high in-situ stress, triggering large deformation of the two sidewalls and even cracking. The failure was mainly induced by large horizontal and lateral pressures. Longitudinal fissures were observed at two sidewalls, and their length basically exceeded the convergence value of 50 mm, whilst the sidewalls deformed inward and the roadway vault was damaged or crushed. The supporting arch deformed upward, thus a peach-shaped or pointed arch under shearing can be observed. If the two sidewalls deform significantly and ultimately spalling occurs, a cavity can be formed between the shotcrete and surrounding rocks. In this case, installed wire mesh can be twisted along with any kind of spray anchor network. The U-steel yieldable arch can also be bent or broken. Ultimately, the spray layer was broken into pieces, and the surrounding rocks were extruded as shown in Fig. 1a and b.

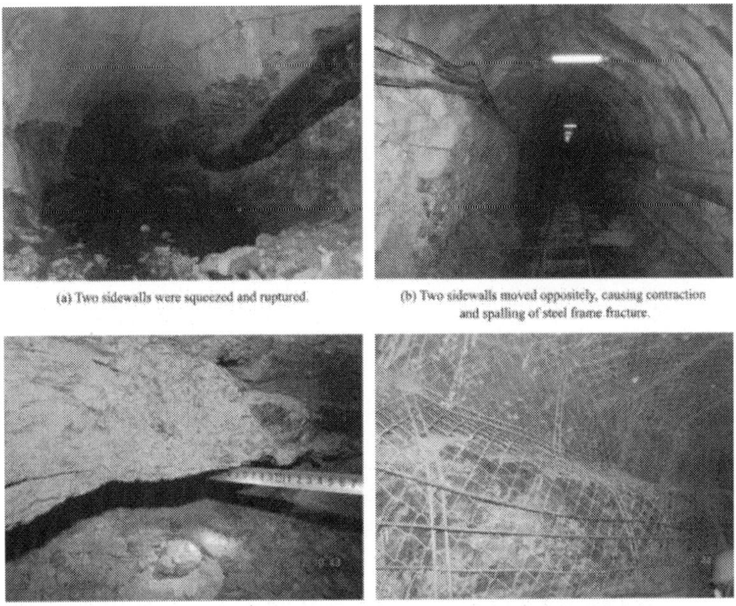

(a) Two sidewalls were squeezed and ruptured. (b) Two sidewalls moved oppositely, causing contraction and spalling of steel frame fracture.

a. (c) Bottom rock mass moved out of roadway. (d) The roadway sidewall heaved.

Figure 1. Deformations of sidewalls of a roadway.

In addition, because rock mass loosening occurred on the two sidewalls, serious collapse at the bottom of roadway could be observed. In the Qujiang coal mine, the collapse in the zone from the upper to the corner of roadways was also serious. Sometimes the entire sidewall of roadway showed strong caving as illustrated in Fig. 1c and d. When serious collapse occurred, a cavity was formed between local anchor support structure and surrounding rocks. In a form of twisting, the steel bars strongly twisted in association with large deformation. Based on field monitoring and observations, the final deformation of the two sidewalls of roadways was approximately 350–520 mm after one month of completion.

2) Most roadway roofs failed with large deformation. Under high in-situ stress, when the roadway roof experienced a large deformation due to overburden stress, bolts, shotcrete, concrete lining, U-steel, and scaffolding tended to yield, causing local or global failure and/or damage. In addition, due to horizontal stress, roadway vault also experienced a large upward displacement. The roadway could not adapt to this large deformation and thus failed, as shown in Fig. 2.

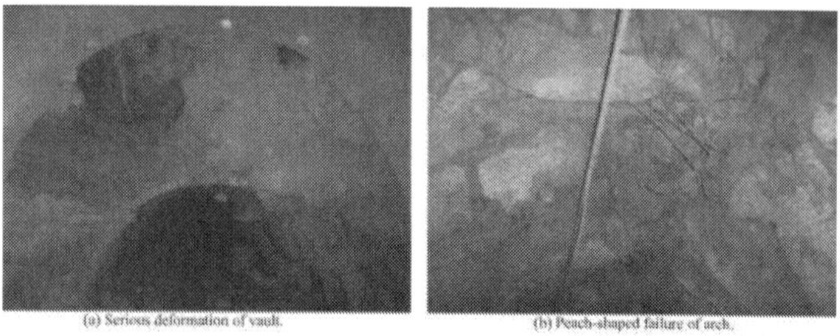

a. (a) Serious deformation of vault. (b) Peach-shaped failure of arch.

Figure 2. Deformation of roadway roof.

3) The floor and bottom corners of roadway were damaged by strong floor heave. Under high in-situ stress, large deformation of surrounding rocks induced by the floor heave can be observed in concrete-uncovered or inverted arch-missing regions. The deformation in these regions is larger than 500 mm or even up to 1000 mm. The larger deformation of surrounding rocks can result in severe uneven deformation of the two sidewalls of the roadway, represented by violent destruction of concrete floor rising from the middle or corners. However, in the roadway with back cover and/or inverted arch, the roadway floor corner will also undergo large deformation under high in-situ stress, leading to bending failure of the whole roadway as shown in Fig. 3.

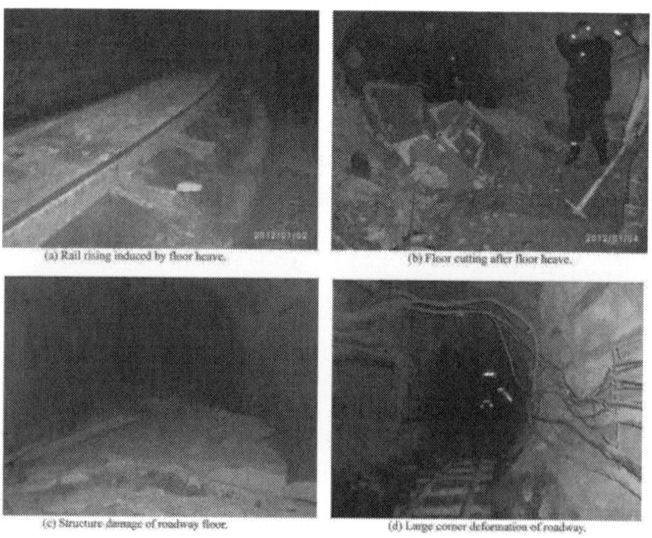

(a) Rail rising induced by floor heave. (b) Floor cutting after floor heave.

(c) Structure damage of roadway floor. (d) Large corner deformation of roadway.

Figure 3. Deformations of roadway floor.

FAILURE MECHANISM ANALYSES OF DEEP ROADWAY

Effect of Soft Rock on Roadway Deformation

The soft rock concept has been proposed and discussed in many studies, e.g. He (2014)and Kanji (2014). Soft rocks have low to extremely low strength, which basically contain clayey minerals of kaolinite, montmorillonite, and illite. Soft rocks under low stress level (e.g. uniaxial compressive stress of 15 MPa) can experience significantly large deformation. Soft rock can thus be regarded as low strength engineered rock, in which the argillaceous rocks are the main component, and the clay minerals can slip and expand under high stresses. Second, due to weak structural face development in rocks, such as small-size faults, interlayer shear surface, bedding plane and joint fissure, rock strength can be significantly reduced. In addition, structural surfaces are usually covered with clay layer, and thus under the action of groundwater, the material of interface will experience softening or argillization, resulting in instability or failure of rock mass. In presence of structural surface, the surrounding rock can be classified as bedded cataclastic structure, cataclastic structure, loose structure, and broken structure (Cai et al., 2013).

The soft rock has the characteristics of plasticity, swelling, disintegration, dispersion, rheology, thixotropy and ion exchange. Unloading effects due to excavation on the rock mass will change the stress state of the surrounding rocks, from three-dimensional (3D) to two-dimensional (2D) because of the release of a directional stress. In order to illustrate the influence of surrounding rocks on roadway stability, some investigations and deformation monitoring of deep roadway in each mining area were carried out, and the cross monitoring method with the vertical line and the laser range instrument was used. The deformation curves and deformation rate curves for different lithology conditions at a depth of 900 m are shown in Fig. 4 and Fig. 5. According to the monitoring results, some conclusions can be drawn as follows:

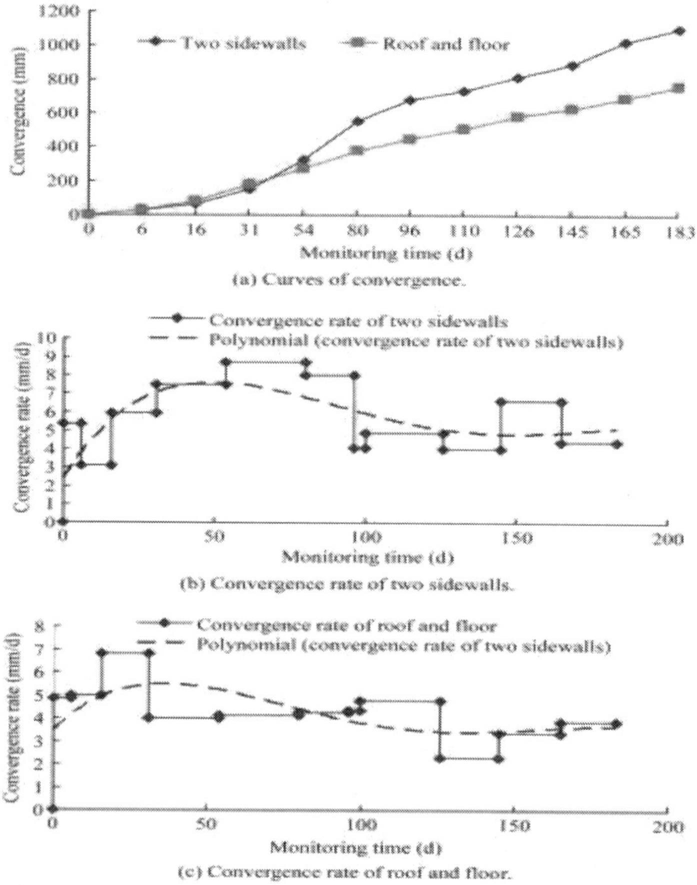

Figure 4. Roadway deformation curves of argillaceous rock mass.

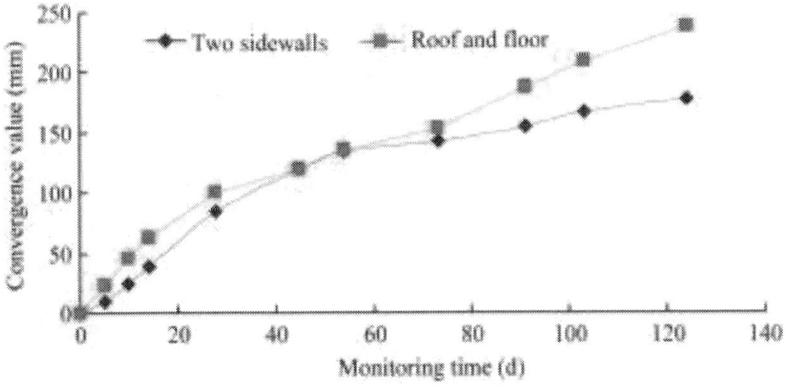

(a) Curves of convergence.

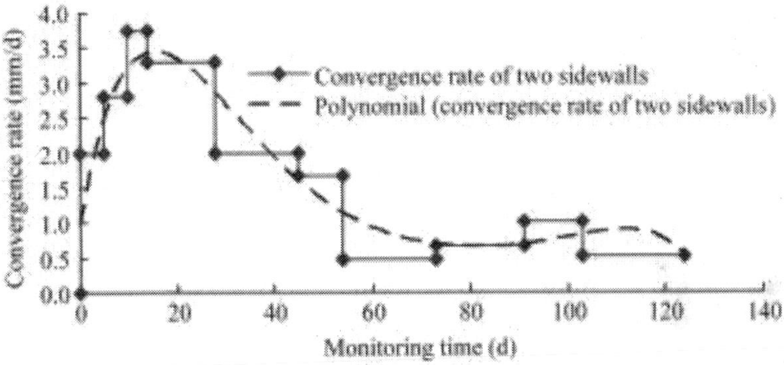

(b) Convergence rate of two sidewalls.

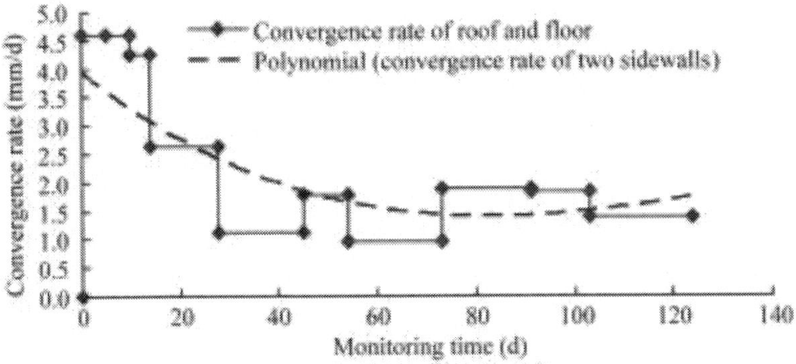

(c) Convergence rate of roof and floor.

Figure 5. Roadway deformation curves of arenaceous rock mass.

1) For rocks with low strength (e.g. argillaceous rock), under the conditions of high stress, the deformation was significantly large at two sidewalls, roof and floor. Fig. 4shows that under the action of high stress, the convergence of two sidewalls reached 1102 mm after 183 days, and that of roof and floor reached 766 mm. After that, the deformation rate was larger than 3 mm/d, and was still unstable at deformation stage.
2) For rocks with high strength (e.g. arenaceous rock) under high stress, the deformation was significantly smaller than that with low strength. It can be seen in Fig. 5 that under the action of high stress, sidewalls convergence reached 178 mm after 124 days, while roof and floor convergence reached 239 mm. After that, the deformation rate was less than 1.5 mm/d, and was stable at deformation stage.

Effects of Cutting type and Weak Interlayer on Roadway Deformation

Effects of Thin Layered Rock on Roadway Deformation

There are basically three kinds of thin layered rocks for mine roadway: horizontal, oblique, and vertical (Fig. 6). In general, when the horizontal stress acts on the layered rock mass, it is very easy to slide because of weak bonding force between layers. For surrounding rocks in deep roadway, due to the large horizontal stress, layered rock will slip to the roadway space when the directions of the layered rock mass and the horizontal stress are similar. In this case, the sidewalls of roadway heaved as shown in Fig. 6a. When the angle between the layered rock mass and the horizontal stress is relatively large, the upper thin layered rock mass on one sidewall will slip more than the lower rock mass, which is induced by the horizontal stress as displayed in Fig. 6b. When the horizontal stress acts in the direction normal to the layered rock mass, the stress state of surrounding rocks is roughly uniform associated with the smallest deformation. However, under high stress conditions, soft rock roadway may present the deformation characteristics as shown in Fig. 6c.

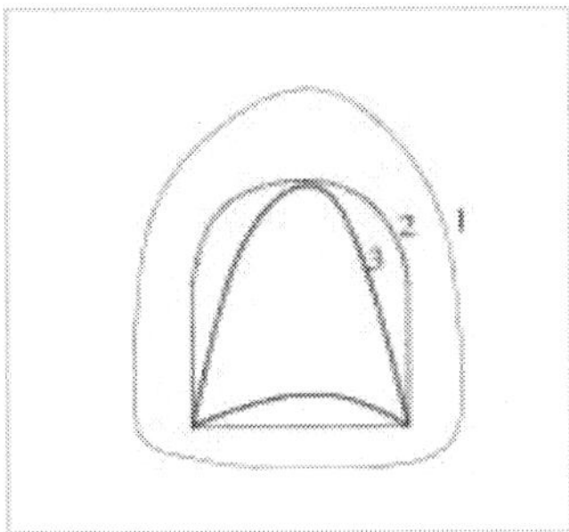

(a) Horizontal thin layered rock blocks and roadway deformation.

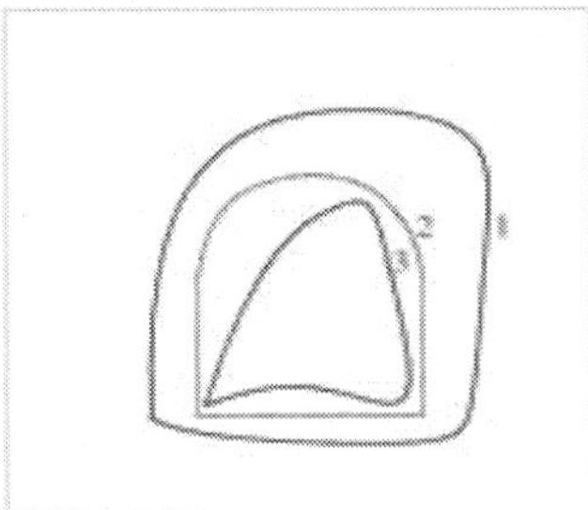

(b) Oblique crossing thin layered rock blocks and roadway deformation.

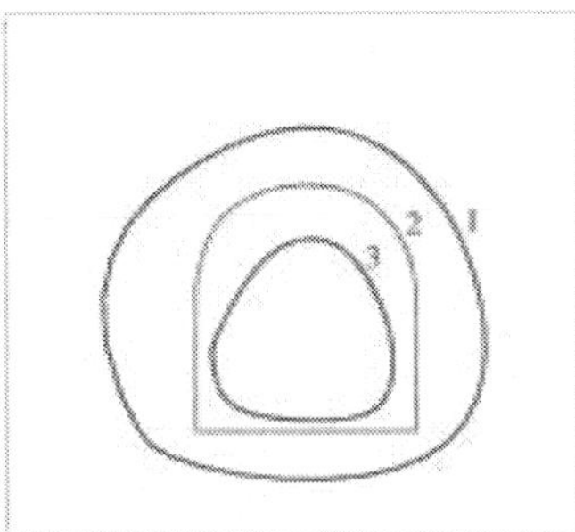

(c) Vertical thin layered rock blocks and roadway deformation.
1—Boundary of broken rock circle; 2—Original roadway;
3—Roadway after deformation

Figure 6. Failure modes of roadway in thin layered rock mass.

Relationship between Rock Block and Deformation of Roadway

Each block cut by joints has a major effect on roadway stability, especially the rock blocks located on the top of roadway as shown in Fig. 7a. The rock mass separated from bedrock will pose directly stress on support structure through gravity. If support strength is weak, rock block falls and free space on the top of roadway occurs. The upper rock on sidewalls of roadway also has a significant effect on roadway stability, which is easy to be separated from the bedrock to produce massive rock. In this regard, upper loose blocks are formed in roadway and are free to side in a form of big blocks. It can be noted that the loose region is considerably large as shown in Fig. 7b.

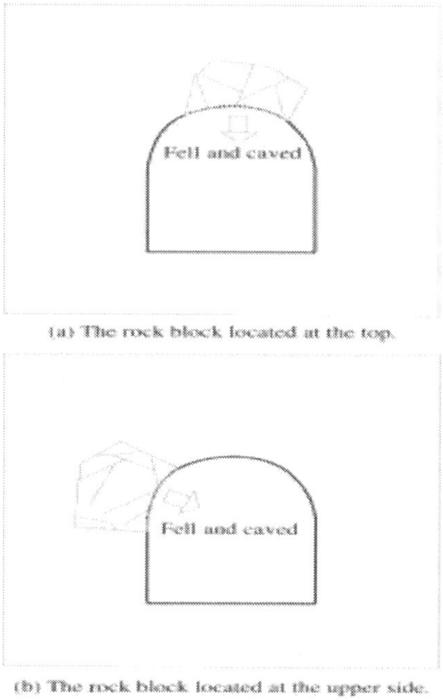

Figure 7. Existing modes of rock block and failure modes of roadway.

When the block rock is separated from bedrock, the roadway surface will be significantly uneven. Small contact area between surrounding rocks and support structure can be observed and stress concentration will be induced consequently under high stress. In this instance, it is most likely to destroy the support structure and the roadway.

Relationship between loose Fractured Rock mass and Deformation of Roadway

In the structurally damaged zone of engineered rocks, typical expansion will be triggered under high stress. The surrounding rocks associated with joints and fissures will gradually expand under the combining actions of tensional, compressive and shear stresses, and thus the joints are dislocated with large displacement. The increase of rock volume induced by rock block rotation and joint expansion are termed as deformation. The fragmentation and expansion of surrounding rocks can lead to excessive deformation or failure of roadway which is categorized as the key block failure around roadway, side shearing failure, combination failure, etc.

In addition, loose fractured rock mass can easily collapse under the action of high stress through stress release. As for the loose rock mass with large porosity, adhesive force among the particles is very low. After excavation of roadway, the surrounding rocks would break very quickly, and form a collapse circle after expanding. Roadway excavation in the original rock mass can break the existing 3D stress state of surrounding rocks, and the stress is redistributed by reducing radial stress to zero, increasing shear stress and generating stress concentration, until a new equilibrium of 3D stress state is reached. In this regard, the stress state of surrounding rocks is changed, and the soft rock strength is significantly reduced. Due to its very low strength, rock fails when the concentrated stress is greater than the strength of surrounding rocks, thus the damaged zone extends from the periphery to the deep fractured zone, which can be called collapse circle of surrounding rocks. Field observations show that the surrounding rocks of roadway have a widespread collapse circle due to excavation-induced stress distribution, which can be also calibrated through laboratory simulation tests. Large collapse circle means great convergence, and therefore it is more difficult to support the roadway after excavation. Generally, the collapse circle is the natural characteristic of soft rock roadway. The thickness of collapse circle of surrounding rocks is the function of stress and strength of surrounding rocks, reflecting the difficulty in rock supporting.

Moreover, for the low strength surrounding rocks, large plastic deformation can be induced under a certain small stress, and it will undergo continuously increasing plastic deformation. The large plastic deformation can result in height reduction of roadway, extrusion of two sidewalls, cross-sectional reduction, etc. It is rather difficult to use low strength supports for the synchronous deformation of surrounding rocks and supports, and therefore large deformation and failure of roadway cannot always be effectively controlled. In this instance, large loose region and large granular pressure of surrounding rocks lead to damage of roadway. When using high-strength supports, the ground pressure imposed on supports would increase with increasing deformation. If the pressure exceeds the limit of support strength, deformation-induced damage to supports would occur. According to the analyses of a variety of roadways, it can be noted that in loose surrounding rocks of roadways, roof damage is caused by dispersed ground pressure, and the damage of sidewalls is caused by shear deformation. After that, the roof falls even if high-performance concrete structures are employed for roadway support due to significantly high shear stress in sidewalls, and a large-scale shear failure would occur in the support structures.

Relationship between Weak Interlayer and Roadway Damage

If weak interlayer goes through the roadway as shown in Fig. 8, shear stress concentration can be generated under high stress, which can lead to larger shear stress than the limit shear strength of structural planes. Thus rock mass structure would slip along the structural planes, and rock mass fails. In particular, soft interlayer would extrude outward under high geo-stress, resulting in local failure or damage of roadway.

Figure 8. Soft weak interlayer through the roadway.

Relationship between Geo-stress Release and Roadway Deformation

Another key factor that can induce large deformation of roadway is the presence of geo-stress. Before excavation of rocks, the geo-stress is in a 3D compression. After excavation, the rocks have a free face, suggesting that the 3D stresses are changed to 2D stresses in association with stress release close to roadway surface. In this case, tensile and shear stresses are generated, which can produce large deformation of surrounding rock or roadway damage.

In order to analyze the relationship between geo-stress release mode and roadway deformation, investigations are made on deep high stress deep roadway, which can be roughly described as follows:

1) Sidewalls of roadway are squeezed by large horizontal stresses, and a sharp peach-shaped roadway can be formed. If the floor support strength is insufficiently high, stress can be released from the floor,

and serious floor heave can be induced subsequently as shown in Fig. 9. The stress redistribution and roadway damage modes are mainly induced by large horizontal stress imposed on two sidewalls, resulting in serious extrusion of sidewalls, and thus a sharp peach-shaped roadway is formed.

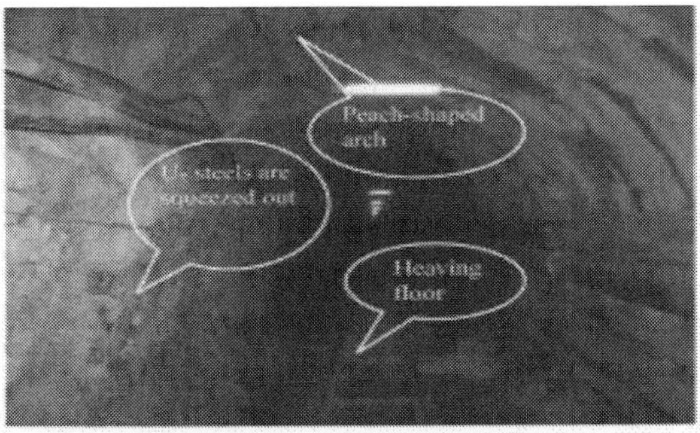

(a) Two sidewalls are squeezed, which is caused by large horizontal stress, forming a sharp peach-shaped roadway and serious floor heave.

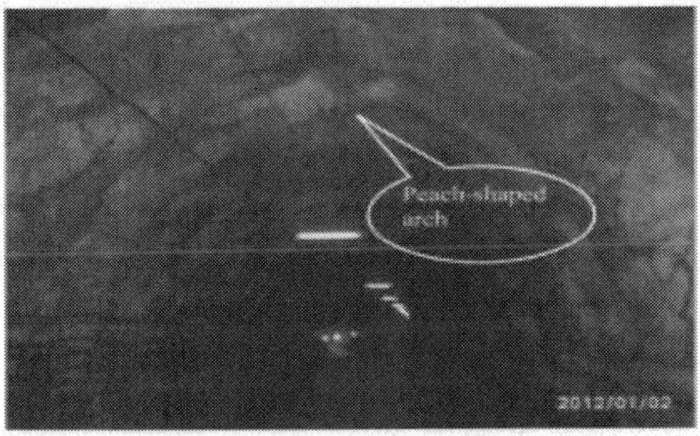

(b) Two sidewalls are squeezed, which is caused by large horizontal stress, forming a sharp peach-shaped roadway.

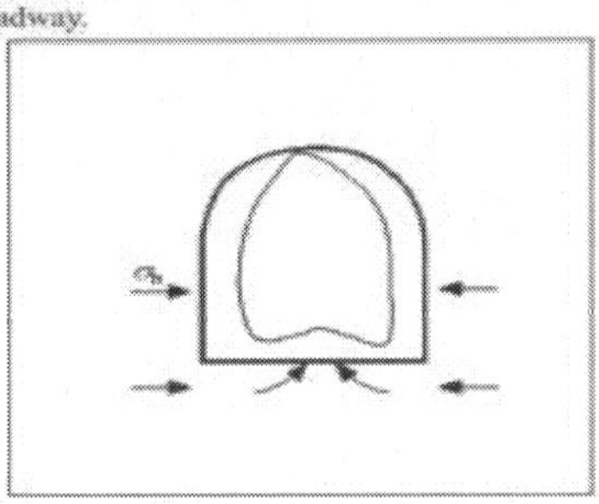

a. (c) Stress transfer and release mode (arrows represent stress release direction).

Figure 9. The actual deformation and stress release mode caused by large horizontal stress.

2) Roadway roof subsidence is caused by large vertical stress; at the same time, stress is transferred to sidewalls and floor. In this case, the stress is released gradually, but the global convergence of roof and floor increases as shown in Fig. 10. Due to the large vertical stress acting on the roof, the roof deforms significantly and subsidence is induced subsequently. In addition, the bottom of the roadway is weakly supported, therefore high stress is released and transferred to sidewalls and floor, suggesting that the floor and sidewalls will deform or extrude more or less.

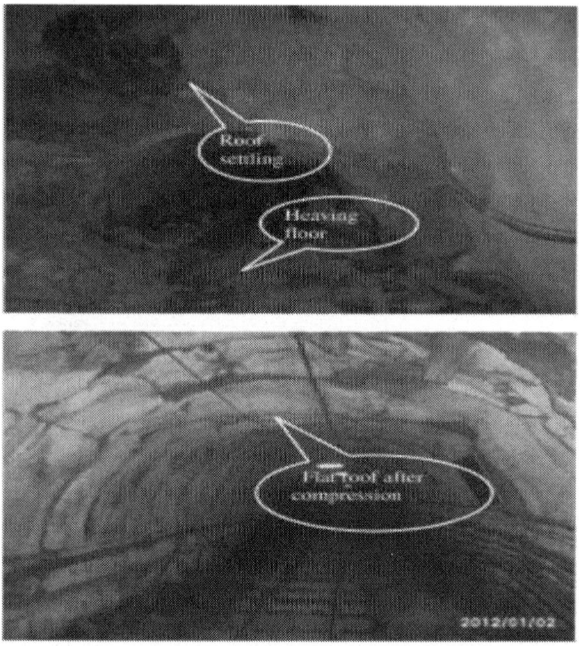

(a) Large convergence of roof and floor caused by vertical stress.

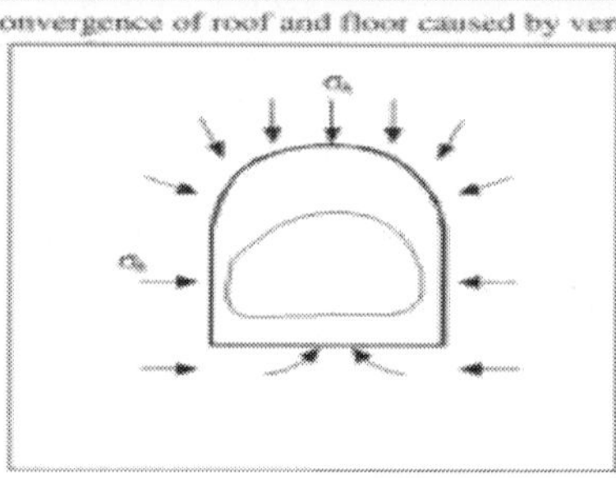

a. (b) Stress transfer and release mode (arrows represent stress release direction).

Figure 10. The actual deformation and stress release mode caused by large vertical stress.

3) Although minor difference may exist between vertical and horizontal stresses, the whole section will also deform under high in-situ stress (>20 MPa). In this condition, stress is basically released to the excavation space as it continuously acts on support structure. The support structure and surrounding rocks deform towards the excavation space, thus some smaller full-section roadways are formed. The deformation and stress release are shown in Fig. 11.

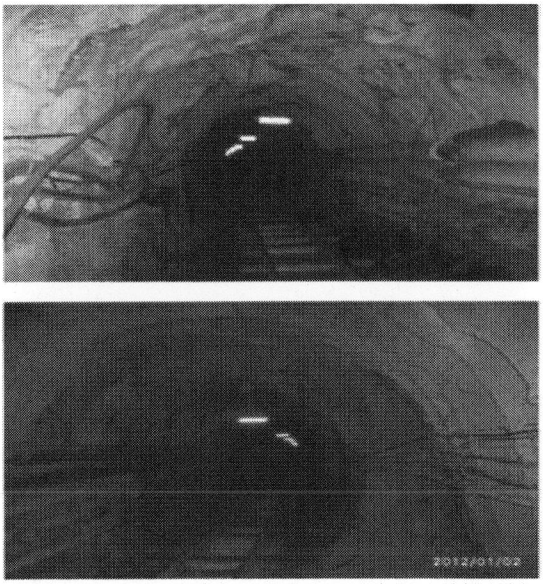

(a) The horizontal and vertical directions are in high stress level, and the whole section shrinks.

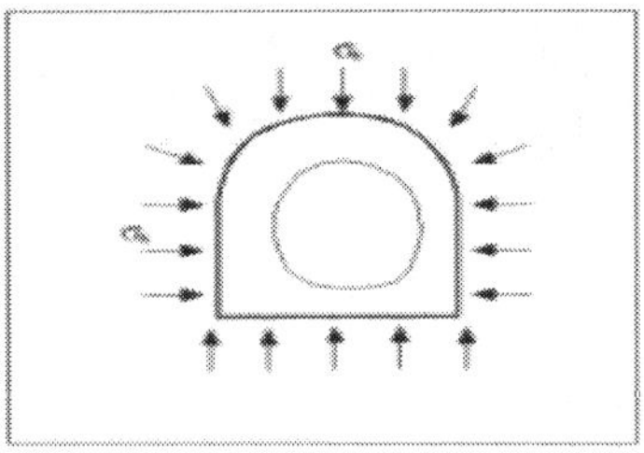

a. (b) Stress transfer and release mode (arrows represent stress release direction).

Figure 11. The actual deformation and stress release mode when the vertical stress is approximately equal to the horizontal stress.

4) When the maximum principal stress intersects with the roadway at a specific angle as shown in Fig. 12, the deformation matches the maximum stress direction. If the section of roadway is symmetric, the

stress release is also symmetric, causing substantive floor heave when the stresses are transferred to the floor.

(a) The horizontal and vertical directions are in high stress level and the whole section shrinks.

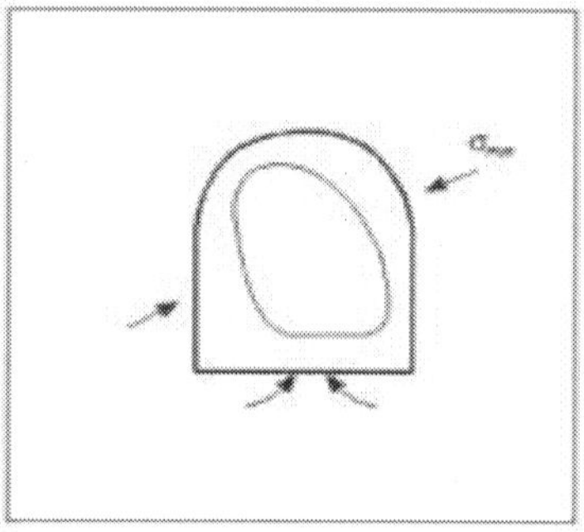

(b) Stress transfer and release mode (arrows represent stress release direction).

Figure 12. The actual deformation and stress release modes when the vertical stress has a certain angle with the horizontal stress.

Relationship between Roadway Deformation and Support Patterns

Bolting-mesh-spray, cable, U-steel, grouting and other supporting patterns are commonly used for surrounding rocks in high stress roadway. However, it is impossible to solve large deformation problem using a single pattern, where the combining support patterns are needed in order to achieve more desirable results. But in actual operation, one support pattern or combining patterns is frequently reported due to various reasons, e.g. lack of construction cost, or the rocks are not supported timely. According to field investigation and related information analysis, different support patterns of roadway in comparison of deformation of surrounding rocks with different characteristics (see Fig. 13) can be specified as follows:

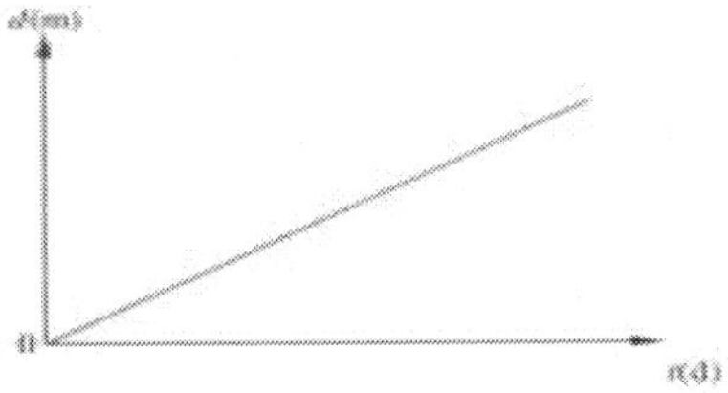

(a) Support pattern of "bolt + wire mesh".

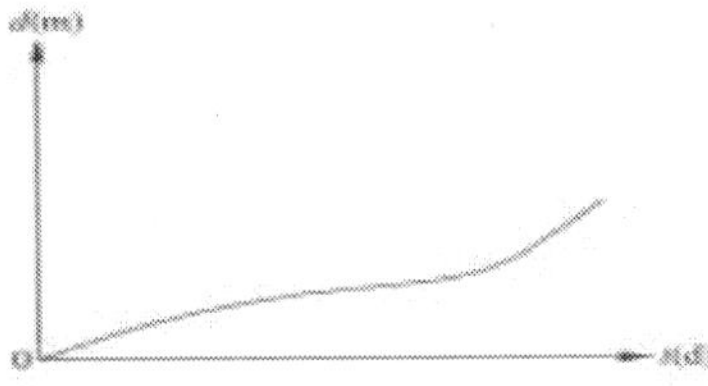

(b) Support pattern of "bolt + wire mesh + shotcrete".

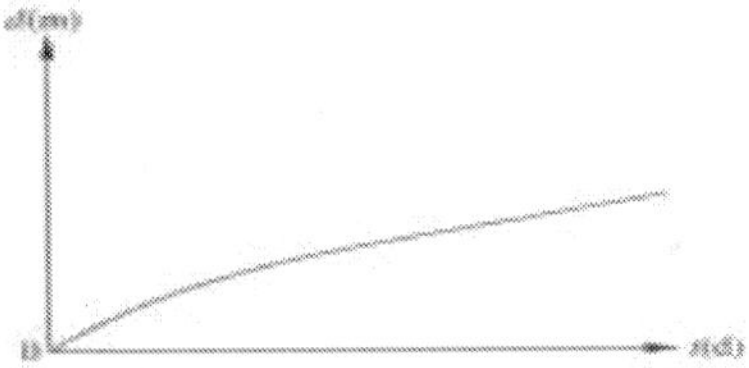

(c) Support pattern of "bolt + wire mesh + shotcrete" + "cable".

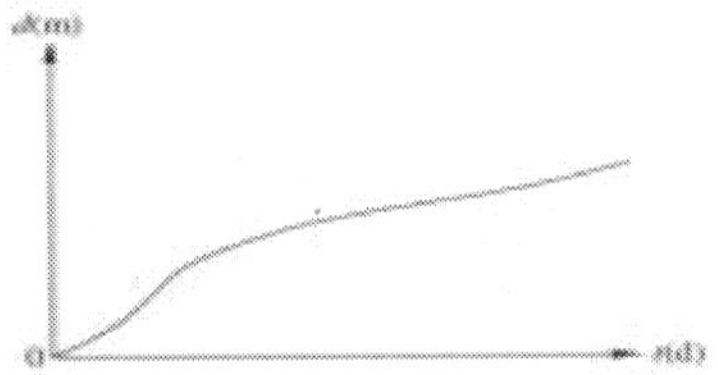

(d) Support pattern of "bolt + metal mesh + U-steel + shotcrete".

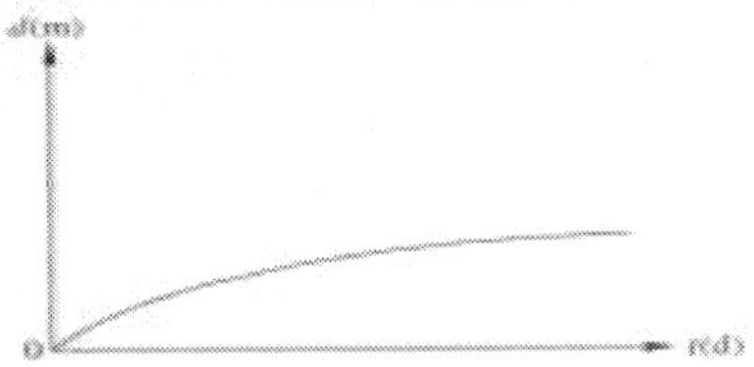

(e) Support pattern of "bolt + wire mesh + shotcrete" + "grouting".

Figure 13. Roadway deformation regulations of different support patterns.

1) When the support pattern of "bolt + wire mesh" is employed for roadway under high stress, the surrounding rocks quickly produce

dilitancy after excavation, and thus large deformation can be induced. Various cracks can also be observed in roadway. Although the "bolt + wire mesh" support pattern is adopted, exposed rocks can be easily weathered or deteriorated because of lack of full-section support of the surrounding rocks. In the roadway containing clay minerals, rock argillization, hydration and swelling can be produced if water is encountered, resulting in side floor heave and roof subsidence. In addition, the support type of roadway can also result in crack extension in soft rock, separated sidewalls, and roof separation. It should be noted that for the "bolt + wire mesh" support, the deformation of surrounding rocks increases linearly till failure, as shown in Fig. 14a.

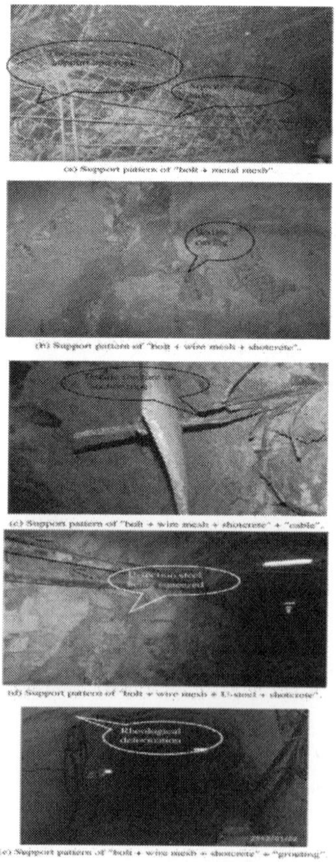

a.

Figure 14. Roadway damage characteristics of different supporting forms.

Therefore, for the support pattern of "bolt + wire mesh" under high stress, the surrounding rocks deform very quickly after roadway excavation in a very short period of time. This support pattern is suitable for temporary support of roadway, but if the secondary support is not carried out in time, surrounding rocks will deform rapidly under the action of high stress till failure of roadway as shown in Fig. 14a.

2) The support pattern of "bolt + wire mesh + shotcrete" is better than the "bolt + metal mesh" support pattern after full-section support (closure) of surrounding rocks is properly considered. In the initial monitoring period, the deformation rate was small; but the loose circle expanded rapidly with the increasing surrounding rock deformation when only bolt support was used. In the later stage, the deformation was in an acceleration stage, resulting in large deformation and instability of roadway. The deformation of roadway is shown in Fig. 14b.

   Generally, roadways using this kind of support pattern are mainly damaged at sidewalls at the beginning. As the damage of sidewalls increases, shotcrete can be separated from roadway surface and its support capability can then be lost. Therefore, the deformation of roadway develops rapidly, and thus rock squeezing, roof subsidence and floor heave occur as shown in Fig. 14b.

3) When the combining support pattern of "bolt + wire mesh + shotcrete" (the first support) and "anchor" (the secondary support) is adopted, the surrounding rocks will remain relatively stable. Evidently, the first support system (bolt + wire mesh + shotcrete) plays a major role in formulating a temporary support, improving initial bearing capacity of the surrounding rocks so as to maintain the integrity of surrounding rocks. In addition, the secondary support (anchor) plays a major role in the permanent support of load-bearing arch structure so as to form a stable structure. Therefore, this support system makes the surrounding rocks more stable, maintaining a constant deformation rate (Fig. 14c). However, due to the high stress state, the deformation would increase with elapsed time, and the anchor or bolt is ultimately broken or cut as shown in Fig. 14c.

4) When the support pattern of "bolt + wire mesh + U-steel + shotcrete" is adopted, the objective is first to allow rock deformation and then to reduce rock deformation. The "bolt + wire mesh" support is flexible,

but the U-steel frame support is rigid. They are integrated to combine "soft" bolt and "rigid" support. Initially, interaction between surrounding rocks and U-steel is weak, and the deformation rate is high. When the deformation increases to a certain value, the pressure of U-steel imposed on surrounding rocks starts to increase and grows larger with increasing deformation to resist the deformation of the surrounding rocks, significantly reducing the deformation rate. However, the U-steel would yield with the increasing deformation if it could not resist the pressure applied by the surrounding rocks. This can result in gradual increase of the deformation rate of the surrounding rocks and eventually lead to instability. The actual deformation of roadway with this support is shown in Fig. 14d.

5) When the support pattern of "bolt + wire mesh + shotcrete" and "grouting" reinforcement is adopted, the grouting can play a major role in support. The grouting can fill the fractures of jointed rock mass and improve the integrity and strength of the surrounding rocks. Thus grouting in association with wire mesh can greatly improve the self-bearing capacity of surrounding rocks. For simplicity, surrounding rocks with the support pattern of this type can remain relatively stable, as shown in Fig. 14e. However, if the surrounding rocks have characteristic of strong rheology, rock deformation will increase under the action of high stress; therefore the roadway can also produce large deformation until instability (Fig. 14e). In addition, the sidewalls, roof and floor can undergo serious deformations when the horizontal stress is high.

Influence of Excavation Disturbance on Roadway Deformation

In mining engineering, underground excavations are complex due to many roadways and chambers close to each other. Thus mining excavation would have an important influence on the stability of surrounding rocks. When blasting is employed, the blasting-induced stress can have damages to the surrounding rocks, leading to the instability of roadway and chambers.

The monitoring curves of haulage roadway and rail roadway before and after excavation in Qujiang coal mine with a mining depth of 886 m are studied, respectively. Two roadways have a parallel spacing of about 20 m, and the haulage roadway is 100 m ahead of rail roadway. The host rock consists of mudstone and silty, black grayish, and thin to medium-thick bedding rocks. The joints are well developed with low strength. The drill-and-blast method is used for full-section excavation.

The influence of excavation disturbance on roadway deformation can be summarized as follows:

1) The drill-and-blasted roadways with a parallel spacing of 20 m have a significantly large impact on pre-existing roadway with a range of around 10–20 m from the heading face center. The impact is characterized by larger displacement and greater deformation rate than other roadways. In this period of time, the distance of the drill-and-blasted roadways with a parallel spacing of 20 m is just about 15 m from the heading face center.
2) The largest disturbance during roadway excavation happens at a distance approximately 25 m away from the working face. The displacement rate is generally greater than 4 mm/d, and is significantly reduced when the advancing is away from the monitoring point. In addition, excavation disturbance to drill-and-blasted roadways is higher than that to adjacent roadways.
3) It is well known that when the two roadways are excavated in close proximity, the excavation-induced disturbance is clearly evident. According to field surveys, when the spacing of roadways is within 30 m (dependent on blasting scale and rock strength), excavation-induced disturbance is large. Deformation of roadway at different distances from the working face is shown in Fig. 15. Excavation zones can be divided into the following ones: excavation damaged zone, major influence zone, minor influence zone, and non-disturbance zone. Thus it can be noted that the excavation damaged and disturbed zones are basically within 30 m from the excavating face.

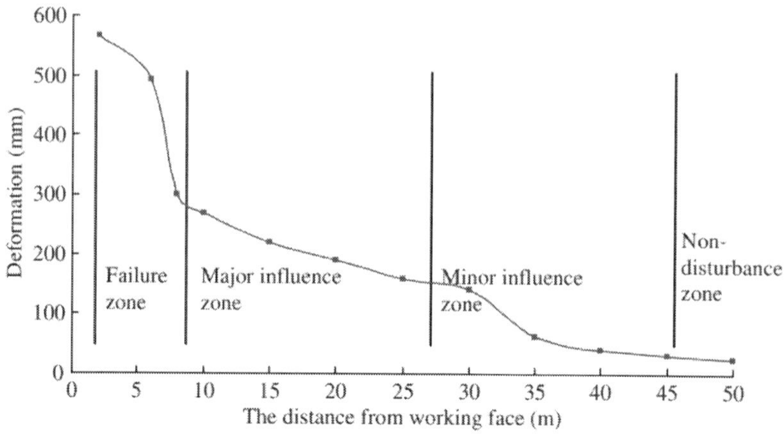

Figure 15. Roadway deformation curve at different distances from working face.

4) Roadway deformation and failure induced by excavation basically have the same characteristics to a certain degree according to field observation. When excavated roadways are adjacent and parallel to pre-existing roadways at nearly the same elevation, one sidewall or arch of roadway will deform, or rock slabbing or spalling will occur.

SUPPORT GUIDELINES AND CONTROL MEASURES OF LARGE DEFORMATION ROADWAYS UNDER HIGH STRESS

Support Guidelines of Roadway

According to roadway damage states and related theories, support guidelines of roadway under high stress can be listed as follows:

1) The primary support stiffness and strength should be significantly improved. This can effectively control the increasing deformation of surrounding rocks, which includes two parts. The first one is the slide along structural surface, crack opening and expansion, which belong to discontinuous deformation. The second is the elastoplastic deformation of rocks before the peak strength is reached, including the global deformation of anchorage zone, which belongs to the

continuous deformation. Because of low strength of structural plane, reasonable roadway supporting pattern can be considered as follows:

i. The initial stiffness and strength of support system are greatly improved to effectively control discontinuous deformation of surrounding rocks. In this regard, the integrity of surrounding rocks can be roughly maintained.

ii. Support system should use suitable technologies, allowing continuous large deformation of surrounding rocks to release high stress.

Compared with traditional support concept of "first using flexible and then rigid" support system, i.e. first to allow and then restrain rock deformation, the support concept in deep and complex roadways is rather different. We should use the principle of "first rigid, then gentle, and finally rigid". The concept can be interpreted that support system should first provide resistance, and then allow rock deformation to a certain degree, and finally limit rock deformation in order to maintain the integrity of surrounding rocks and keep the strength of surrounding rocks as possible.

2) The support stiffness should be improved to form high bearing capacity of prestressed structure at the earliest. Prestressed bolt support mainly plays a role in controlling sliding and moving of surrounding rocks of anchorage zone and reducing crack initiation and propagation. In this instance, the surrounding rocks are in a state of compression to reduce tensile-and-shear failure and bending deformation of rocks. The prestressed bearing structure of high rigidity should be kept in the anchorage zone in order to prevent separation of layers outside the anchorage zone and to improve stress distribution in deep rocks.

3) The bolt plate, steel belt and wire mesh can be employed to increase prestressed area and to support surrounding rocks. According to site-specific conditions of roadway, the key of the support design is to obtain an effective prestress loss when determining reasonable scheme. The scope of one single bolt prestress is very limited, thus the plate, steel belt and wire mesh can be used to transfer stress to far-field surrounding rocks. For the roadway surface, even a small magnitude of support force

can significantly keep the integrity of the roof-and-rock structure and reduce the deformation of surrounding rocks.

4) Engineers should adopt the support pattern of “bolt and cable” in association with high prestress, high strength and high stiffness, to effectively control rock deformation and failure and to avoid secondary support or unexpected roadway maintenance.

5) If necessary, grouting can be used to improve the strength and integrity of broken rock mass so as to keep the integrity of loose zone.

Control Measures

Based on the above analyses, the core idea of this kind of roadway support is to improve internal rock strength as possible, and prevent concentrated stress of surrounding rocks transferring to other zones. The overall support technology of roadway can thus be determined by strengthening the support system incrementally. Therefore, according to the deformation characteristics of surrounding rocks, a comprehensive support method of “bolt + wire mesh + U-steel arch bracket + shotcrete + grouting and cable” can be used for the whole section.

The following scheme can be applied according to site-specific conditions in different roadways, thus the parameters and construction method will also differ:

1) Primary support: bolt, wire mesh, U-steel arch bracket, shotcrete, and ladder beam.
2) Secondary support: two-step grouting, anchor and cable beam with leg lock.
3) Integrated technology of floor reinforcement: lock legs in middle position of each U-steel frame by using lap type cable, and the U-steel bracket to maintain the integration of the whole roadway section.

CASE STUDIES

Taking the −850 m level roadway in Qujiang mine for example, the main haulage roadway extends to a new excavation section, which is mainly composed of weathered layer of siltstone. The lithology of soft rocks is black, thin to medium-thick bedding associated with thin mudstone and fine sandstone belt. It contains a small amount of siderite concretions, and joints are well developed. From the monitored deformation data of pre-existing roadways, it is known that large deformation of surrounding rocks could last for a long time. Thus the roadway was repaired for many times, and the maintenance work was very large. The support and repair cost reached more than 20,000 RMB/m. In order to address the problem, a new support scheme should be considered for these roadways, i.e. a comprehensive support method of "bolt + wire mesh + U-steel arch bracket + shotcrete + grouting" + "cable".

(1). Bolt Support Parameters

Thread bar bolt without longitudinal rib of the size of ϕ22 mm × 2200 mm associated with BHRB500 rebar without longitudinal rib was used for this project. Three K2350 resin anchoring agents were used for each bolt with anchoring force not less than 70 kN. The spacing was 700 mm × 700 mm as shown in Fig. 16a. Full-section was considered with wire mesh and steel ladder beam. The wire mesh has a diameter of 6 mm and the spacing of 100 mm × 100 mm, which was determined according to specifications for the wire mesh of the size of 1000 mm × 800 mm. Wire mesh junction must be bolted, and each neighboring lap length is no less than 100 mm. The steel ladder beam was welded by round steel with the diameter of 12 mm.

(2). U-steel Bracket

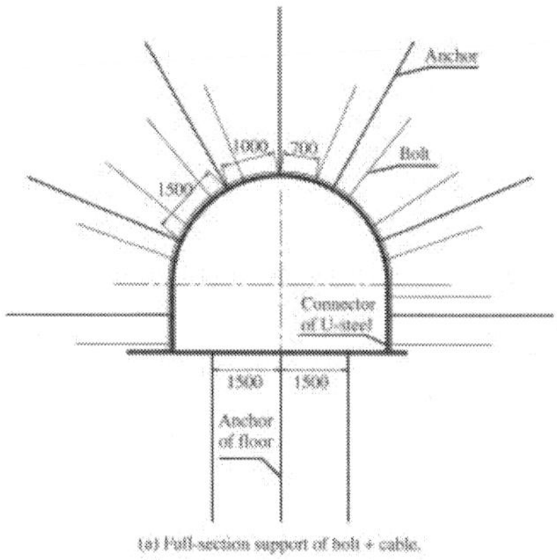

(a) Full-section support of bolt + cable.

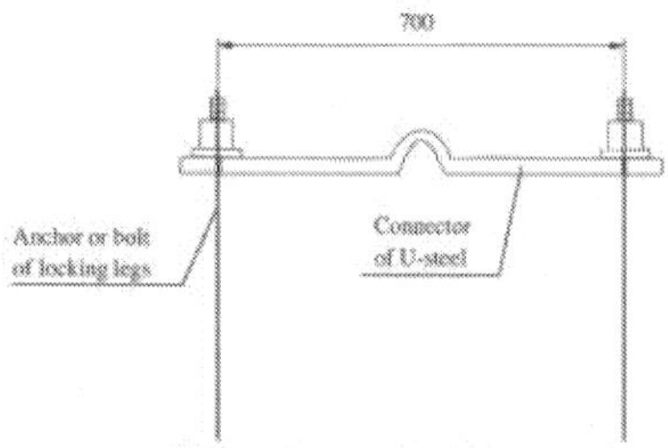

(b) Device of lap type clip lock leg.

Figure 16. Bolt + cable support scheme of east main roadway (unit: mm).

U29 type steel bracket was employed in this project with the spacing of 700 mm. The U-steel bracket has the advantages of high resistance, shrinkage and strong protection surface, but as a kind of passive support it is not able to control and adapt to the complex deformation characteristics of roadway. Thus the weakness of the structure should be considered and reinforced. To lock legs in the waist of each U-steel frame, using lap type clip can form continuous constraints to sidewalls (Fig. 16b).

The U-steel bracket of leg lock mainly has the following two roles: (i) similar action to pull bar by connecting the brackets to improve the integrity and stability of the whole support system; and (ii) insertion into the stable surrounding rocks to increase the capacity to resist lateral pressure, which can effectively control the inward roadside deformation.

(3). Cable Support

Cable with steel strand was used with the diameter of 21.6 mm, and the length of 6.3 m, as shown in Fig. 16a. The spacing is 1000 mm × 1500 mm. End resin anchorage was used with the anchorage length of 1600 mm. Four K2350 resin anchoring agents were used for each cable, and the anchoring force of each cable is no less than 120 kN. The size of two anchor plates with two pad stacks was 350 mm × 350 mm × 10 mm and 150 mm × 150 mm × 10 mm, respectively. The large one was placed in an upper position and smaller one below. The cable clip used to lock leg was considered with pressed steel thickness of 10–15 mm and width of 15–20 mm. A single cable clip was used to lock a single U-steel, and holes were drilled at the two cable (bolt) ends.

(4). Grouting Scheme and Parameters

The grouting is divided into two types, i.e. shallow-hole grouting (2 m) and deep-hole grouting (5 m). For the shallow-hole grouting, the spacing of the arch and sidewalls is 2000 mm × 1500 mm, and the floor spacing is 1200 mm × 1200 mm, as shown inFig. 17a. For the deep-hole grouting, the spacing of arch and sidewalls is 2000 mm × 1500 mm and the floor spacing is 2000 m × 1200 m, as shown in Fig. 17b. When grouting, the first step is to conduct shallow-hole grouting, and then deep-hole grouting. A certain time interval between two steps of grouting should be considered, i.e. the deep-hole grouting can be implemented only when the shallow-hole grouting strength reaches 80% of design strength.

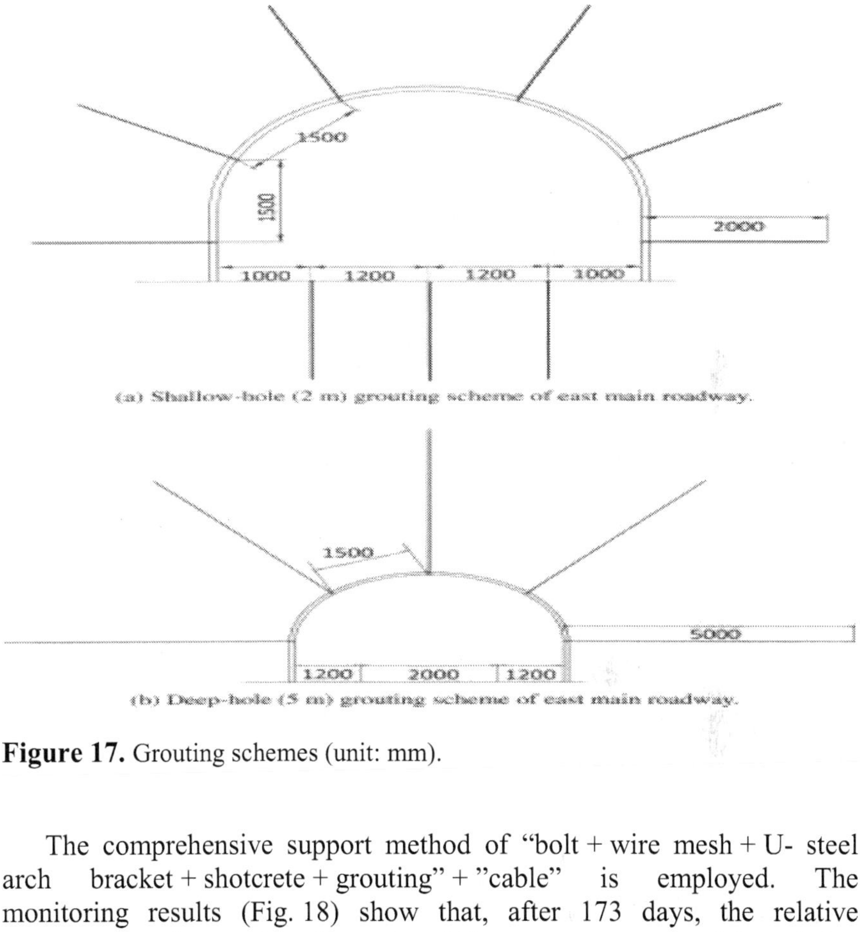

Figure 17. Grouting schemes (unit: mm).

The comprehensive support method of "bolt + wire mesh + U- steel arch bracket + shotcrete + grouting" + "cable" is employed. The monitoring results (Fig. 18) show that, after 173 days, the relative convergence of two sidewalls reaches 208 mm on average, and the relative convergence of roof and floor reaches 448 mm. In addition, for the new roadways without using the comprehensive support method, the relative convergence of two sidewalls reaches 723 mm on average, and the relative convergence of roof and floor reaches 1201 mm after 143 days. It can be noted the that total deformation of the surrounding rocks in test section is much smaller than that of the new roadway without using the comprehensive support scheme, which has better effects on high stress soft rock roadway with large deformations.

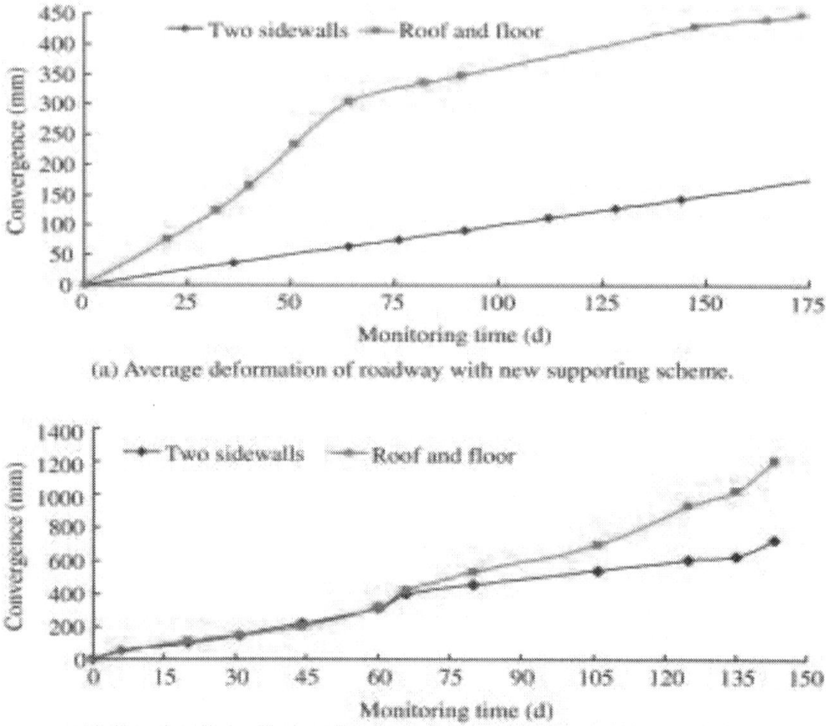

(a) Average deformation of roadway with new supporting scheme.

(b) Average deformation of roadway without new supporting scheme.

Figure 18. Deformation monitoring curves of roadway.

CONCLUSIONS

1) Field investigation shows that the deformation of high stress soft surrounding rocks can be interpreted by large horizontal lateral pressure, which induces two sidewalls to be squeezed significantly. The large deformation of sidewalls can induce serious cracking, and thus strong caving and hollowing can occur. The high stress of roadway can lead to bottom bulge deformation and bending failure of the whole section, if not adequately supported.
2) Soft rock plays an important role in the stability of high stress roadways. On one hand, unloading effect of rock mass due to excavation can change the stress state of surrounding rocks, from 3D to 2D. On the other hand, large deformation occurs for surrounding rocks of low strength under high stress. Therefore, rocks with low

strength (e.g. argillaceous rock roadway) under the conditions of high stress can deform at sidewalls and/or roof and floor.

3) Different weak interlayers or structures have different impacts on deformation and failure modes of roadway. For thin bedding within the roadway, the damage is mainly caused by the thin bedding occurrence. The damage between block rock and roadway can be attributed to the main performance of the block separated from the host rock, making the roadway surface significantly uneven. The small contact area between surrounding rocks and support structure can produce stress concentration likely to destroy the support structure and the roadway. Loose and fractured rocks under the action of high stress can burst easily, forming a large rupture area. In other words, if there is a weak interlayer crossing the roadway, it is prone to generate shear stress under the action of high stress. The accumulated shear stress can be larger than rock shear strength and rock fails consequently.
4) One key factor likely to cause roadway deformation is the presence of geo-stress. Two sidewalls of roadway can be squeezed by large horizontal stress, and a peach-shaped roadway can be formed. Roadway roof subsidence is caused by large vertical stress. At the same time, stresses are transferred to sidewalls and floor gradually, increasing the convergence of roof and floor. Although a minor difference may exist between the vertical stress and horizontal stress, the whole section shrinkage of roadway can be induced in high stress state. When the maximum principal stress intersects with the roadway at a certain angle, a certain magnitude of stress can also be transferred to the floor, causing significant floor heave.
5) For soft rocks in high stress condition, comprehensive supporting patterns should be used. Based on field investigations and theoretical studies, different support patterns can be employed to control the deformation of roadway associated with different rock characteristics.
6) Roadway deformation and failure caused by excavation have a close relation with the elevation difference of adjacent roadways. According to field observations, for parallel roadways, large deformation of side walls or arch of roadway can be observed in the vicinity of the disturbance zone, which can also induce the whole section to deform seriously. Based on the surrounding rock deformation characteristics,

excavation zones would be divided into excavation damaged zone, major influence zone, minor influence zone, and non-disturbance zone.

7) Based on deformation characteristics of high stress soft rock roadway, the support principles, mitigations, and design methods are proposed. In this regard, the proposed comprehensive support method of "bolt + wire mesh + U-steel arch bracket + shotcrete + grouting" + "cable" was employed in east section of roadway of −850 m level in Qujiang coal mine. The monitoring results show that after 173-day monitoring, the average convergence of sidewalls reached 208 mm, the average relative convergence of roof and floor reached 448 mm, suggesting that this kind of supporting method for controlling large deformations of high stress soft rock roadway is effective.

CONFLICT OF INTEREST

The authors wish to confirm that there are no known conflicts of interest associated with this publication and there has been no significant financial support for this work that could have influenced its outcome.

ACKNOWLEDGMENTS

This work was financially supported by the National Natural Science Foundation of China(Grant Nos. 51434006, 51374105, 51374106) and Scientific Research Fund of Hunan Provincial Education Department (Grant No. 14A045).

REFERENCES

1. Bai JB, Wang XY, Yao Z. Study of coupling support in soft rock roadway under high stress. Journal of China University of Mining and Technology 2007;36(4):421e5 (in Chinesse).
2. Cai MF, He MC, Liu DY. Rock mechanics and engineering. 2nd ed. Beijing: Science Press; 2013 (in Chinese).

3. Chen X, Guo HY, Zhao P, Peng X, Wang SZ. Numerical modeling of large deformation and nonlinear frictional contact of excavation boundary of deep soft rock tunnel. Journal of Rock Mechanics and Geotechnical Engineering 2011;3(Supp.): 421e8.
4. Gonzalez-Nicieza C, Alvarez-Virgil AE, Gonzalez-Palacio C, Menendez-Diaz A. In- fluence of the depth and shape of a tunnel in the application of the convergence-confinement method. Tunnelling and Underground Space Technology 2008;23(1):25e37.
5. Hayano K, Matsmoto M, Tatsuoka F, Koseki J. Study of triaxial creep testing method and model for creep deformation on sedimentary soft rocks. In: Proceedings of the 29th Symposium of Rock Mechanics. Tokyo, Japan: Committee of Rock Mechanics, Japan Society of Civil Engineers (JSCE); 1999. p. 8e14.
6. He MC, Guo ZB, Ren AW, Hu YG. Deep soft rock supporting technology of the repair project for transport tunnel in Liuhai coal mine. Chinese Journal of Geotechnical Engineering 2005;27(9):977e80 (in Chinese).
7. He MC, Li GF, Wang J, Cai J. Study on supporting design for large area serious roof caving of deep soft rock roadway in Xing'an coal mine. Chinese Journal of Rock Mechanics and Engineering 2007;26(5):959e64 (in Chinese).
8. He MC, Yuan Y, Wang XL, Wu ZQ, Liu C, Jiang YL. Control technology for large deformation of mesozoic compound soft rock in Xinjiang and its application. Chinese Journal of Rock Mechanics and Engineering 2013;32(3):433e41 (in Chinese).
9. He MC. Latest progress of soft rock mechanics and engineering in China. Journal of Rock Mechanics and Geotechnical Engineering 2014;6(3):165e79.
10. Kanji MA. Critical issues in soft rocks. Journal of Rock Mechanics and Geotechnical Engineering 2014;6(3):186e95. Snupárek R, Kone cný P. Stability of roadways in coalmines alias rock mechanics in practice. Journal of Rock Mechanics and Geotechnical Engineering 2010;2(3): 281e8.
11. Verma AK, Singh TN. Assessment of tunnel instabilityda numerical approach. Arabian Journal of Geosciences 2010;3(2):181e92.
12. Wang WJ, Peng G, Huang J. Research on high-strength coupling support technology of high stress extremely soft rock roadway. Journal of China Coal Society 2011;36(2):223e8 (in Chinese).
13. Xie WB, Jing SG, Wang T, Ren YK, Zhang N. Structural stability of U-steel support and its control technology. Chinese Journal of Rock Mechanics and Engineering 2010;29(Supp. 2):3743e8 (in Chinese).
14. Yu WJ, Gao Q, Zhu CQ. Study of strength theory and application of overlap arch bearing body for deep soft surrounding rock. Chinese Journal of Rock Mechanics and Engineering 2010;29(10):2134e42 (in Chinese).

15. Yu WJ, Wang WJ, Huang WZ, Wu H. Deformation mechanism and rework control technology of high stress and soft rock roadway. Journal of China Coal Society 2014;39(4):614e23 (in Chinese).

CITATION

Weijian Yu, Weijun Wang, Xinyuan Chen, Shaohua Du, Field investigations of high stress soft surrounding rocks and deformation control, Journal of Rock Mechanics and Geotechnical Engineering, Volume 7, Issue 4, August 2015, Pages 421-433, ISSN 1674-7755, http://dx.doi.org/10.1016/j.jrmge.2015.03.014.

CHAPTER 5

Geotechnical Distinction of Landslides Induced by Near-Field Earthquakes in Niigata, Japan

Hirofumi Toyota and Susumu Takada

Department of Civil and Environmental Engineering, Nagaoka University of Technology, Nagaoka, Niigata 940-2118, Japan

ABSTRACT

Landslides triggered by near-field earthquakes with epicentres directly beneath towns have attracted intense attention since the 2004 Mid-Niigata (Niigata-ken Chuetsu) Earthquake. Hilly and mountainous areas sustained heavy damage. Social problems developed when many towns became isolated because landslides cut off traffic and public service lifelines. Soil from landslides closed river channels and formed natural dams. The natural dams submerged some towns. Emergency measures were undertaken promptly to prevent debris flows caused by natural dam breaks. Subsequently, the 2007 Mid-Niigata Offshore (Niigata-ken Chuetsu-oki) Earthquake and the 2011 Northern Nagano Earthquake struck the Niigata region. Landslides triggered by those earthquakes differed in terms of their number, scale, and location. Therefore, characteristics of the landslide sites of the respective earthquakes were examined to ascertain their topographical and geological features. Furthermore, differences in groundwater level and damage related to compound disasters were explained for discussion of the stability progress of damaged slopes.

INTRODUCTION

Niigata prefecture is located in western and central Japan (Figure 1). In the midlands of Niigata, the Chuetsu region, various natural disasters have frequently struck in recent years. At 17:56 on 23 October, 2004, the Mid-Niigata (Niigata-ken Chuetsu) Earthquake, the main tremor of which had a magnitude of 6.8, struck central Niigata-ken (Chuetsu area) and severely damaged the infrastructure of hilly and mountainous areas including Kawaguchi town, Ojiya city, Nagaoka city, and their environs. Numerous landslides occurred especially in the "Yamakoshi area" (Figure 1). That village was isolated by the cutting of all roads and infrastructural lifelines. The epicentral thrust-fault earthquake had a hypocentre of about 13 km depth, with characteristic frequent strong aftershocks that engendered further damage. In addition, rainfall of more than 100 mm was recorded from typhoon number 23, which had passed through the Chuetsu region three days priorly. Data from the Japan Meteorological Agency (JMA) indicated that the daily rainfall of 21 October 2004 at Nagaoka city reached 115 mm. Under those circumstances, more than 3,000 landslides occurred in the hilly area close to the seismic centre during the earthquake. The Japan Society of Civil Engineers [1] and Toyota et al. [2] reported specific damage caused by the disaster. Moreover, Tsukamoto et al. [3] and Rathje et al. [4] estimated the ground movement caused by landslides during the earthquake using investigation or remote sensing data. Geological and geomorphological features and special distributions of landslides in this area were reported by Chigira and Yagi [5] and Wang et al. [6]. Analyses of landslides including discontinuous dip layer were conducted by Onoue et al. [7] and Deng et al. [8]. They assumed that the landslides had been triggered by increase of pore water pressure in seam layers [9] and analysed the displacement using cyclic loading test results or Newmark's method [10].

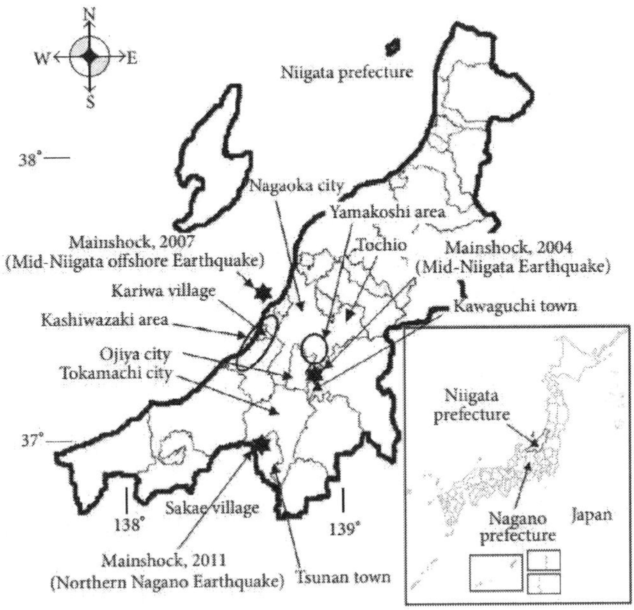

Figure 1. Map of Niigata prefecture.

Heavy snow fell during the two winters following the Mid-Niigata Earthquake. Yearly snowfall is usually about 400 cm in Nagaoka, but it reached nearly 700 cm per year in each of the two years following the earthquake. The snow in one year was the heaviest in 19 years. The collapsed slopes exhibited interesting behaviour during snow and snow-melting seasons. Actually, this region has been notorious as a tertiary type landslide area. Landslides have frequently occurred in April and May: the snow-melting season.

Just three years after the Mid-Niigata Earthquake, at 10:30 on 16 July 2007, the Mid-Niigata Offshore (Niigata-ken Chuetsu-oki) Earthquake, the main tremor of which had a magnitude of 6.8, struck offshore about 30 km northwest of the epicentre of the Mid-Niigata Earthquake. The earthquake was caused by a thrust fault with a hypocentre of about 17 km depth. The JMA Seismic Intensity was recorded as "upper 6" in Kashiwazaki city, Nagaoka city, and Kariwa village. Major landslides were concentrated almost entirely along the "Kashiwazaki area" coastline (Figure 1). Earthquake damage was reported by the JGS [11], Onoue and Toyota [12], and Toyota and Onoue [13].

At 3:59 on 12 March 2011, immediately after the 2011 off the Pacific coast of Tohoku Earthquake (14:46 on 11 March 2011) [14], another earthquake (designated as the Northern Nagano Earthquake) occurred at the border of Nagano and Niigata prefectures [15]. Its main tremor had a magnitude of 6.7. The earthquake, caused by a thrust fault, had a hypocentre of about 8 km depth. The JMA Seismic Intensity was recorded as "upper 6" in Sakae village (Figure 1). Some landslides occurred in Sakae village, Tokamachi town, and Tsunan town under the remaining snow conditions.

For those successive earthquakes which occurred in close proximity, the landslide types are investigated for each earthquake and are compared considering the importance of groundwater levels on landslides. Moreover, the actual conditions of compound natural disasters were examined in the Chuetsu area to assess the compounded damage of these successive natural disasters.

GEOLOGICAL AND GEOMORPHOLOGICAL FEATURES

In the Chuetsu region, thick alluvium covers plains; hilly areas are composed mainly of soft mudstone of quaternary and tertiary deposits (Figure 2). Quaternary deposits are new strata formed from 2 million years ago. Tertiary deposits are geological structures formed between 24 million and 2 million years ago. The geomorphology formed by folding presents a prominent landslide area in this region.

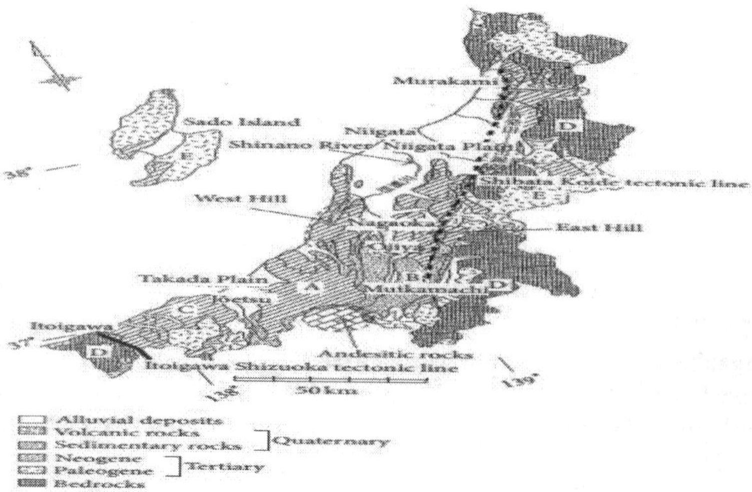

Figure 2. Geological map of Niigata [16].

The Chuetsu region was originally notorious as a tertiary type landslide area [16]. This area has been struck by many natural disasters in recent years. In the Chuetsu area, including Nagaoka and Ojiya (Figure 1), thick alluvium covers the Niigata plain created by the Shinano River. Hilly areas are composed mainly of soft mudstone of quaternary and tertiary deposits. This region has been compressed along the northwest-southeast axis because of crustal movements. Its folded mountains present prominent landslide configurations such as the cuesta landform. Earth-related disasters are generally concentrated in the "West Hills" and "East Hills" (Figure 2). The latter region includes a catchment area dotted with many ponds and rice terraces.

Tertiary mudstone is distributed in the Tokamachi and Tsunan areas, which are near the Northern Nagano Earthquake epicentre. This geology resembles that of the East Hills and is also notorious as a tertiary type landslide area. Nevertheless, it is interesting that Andesitic rocks mainly compose mountainous areas in the northern part of Nagano when crossing the prefectural border separating Niigata and Nagano (Figure 2).

CHARACTERISTICS OF LANDSLIDES

The Mid-Niigata Earthquake

Old Landslides

"Tanesuhara," "Mushigame," and "Asahikawa" in Yamakoshi area have been described as representative designated landslide areas. Figure 3 presents a landslide map of those areas during the Mid-Niigata Earthquake provided by the Geospatial Information Authority of Japan (GSI). River-clogging landslides along the Imo River during the earthquake are the "Terano," "Nampei," "Naranoki," "Higashi-takezawa," and "Junidaira" landslides from upstream (Figure 3). Among those landslides, Terano and Higashi-takezawa were large-scale landslides, which required urgent countermeasures against river clogging.

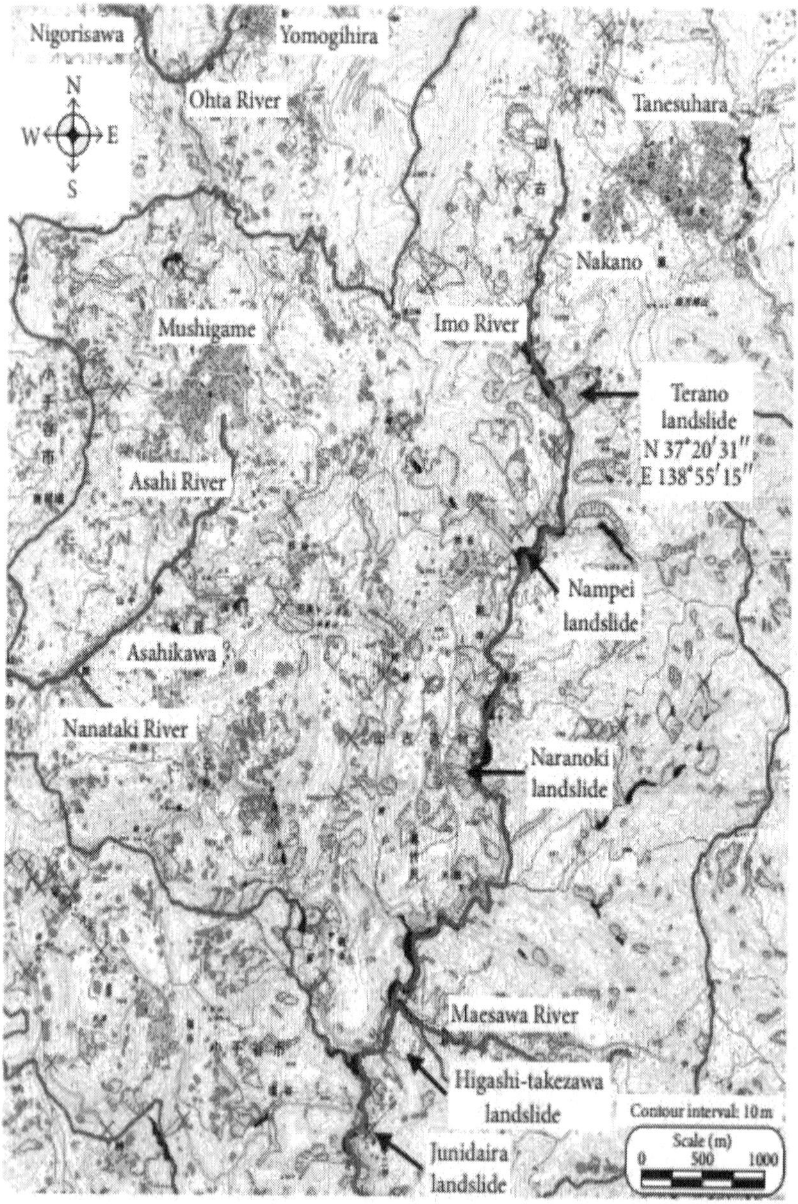

Figure 3. Disaster map around Yamakoshi area during Mid-Niigata Earthquake (provided by GSI).

Table 1 presents the number of landslides occurring during 1949–2002 around Yamakoshi area [16]. Many landslides occurred in the Asahi River basin, extending from Mushigame, and in the Ohta River, which has

Yomogihira and Nigorisawa as designated landslide areas. Although 17 landslides occurred in the Imo River basin, almost all occurred in the Tanesuhara or Nakano landslide area located upstream. Landslides were reportedly only a few in Nampei, Higashi-takezawa, or Junidaira, where natural dams were created as a result of the earthquake. When classified by region (Table 1(b)) from most to least numerous, they are Asahikawa, Tanesuhara, and Mushigame.

Table 1. Number of landslides during 1949–2002.

(a) Classified by basin

	Number of landslides
Asahi River	29
Ohta River	19
Nanataki River	19
Imo River	17

(b) Classified by region

	Number of landslides
Asahikawa	32
Tanesuhara	15
Mushigame	7

The history of the former Yamakoshi village is well recorded [17]. Its history from 1700 indicates Tanesuhara as the site of the greatest recorded earth-flow disaster. Table 2 shows records of landslides in Tanesuhara. The landslides are only recorded in Tanesuhara because no such large villages existed in Yamakoshi area. Although landslides in prehistoric times are unknown, the first landslide recorded in Tanesuhara occurred in 1824. An exogenous factor causing landslides is snow-melt water in early spring. At that time, slopes were destroyed and two large clogging ponds were created in the Imo River. About a century after that event, a landslide occurred during the snow-melting season at almost the same place, thereby forming a natural dam. During the Mid-Niigata Earthquake, which occurred about 80 years after the previous event, the river-clogging landslide was broken at Terano near Nakano. The decisive difference is that the expected exogenous factor was not snow-melt runoff, but an earthquake. As described above, river clogging caused by landslides is a common phenomenon in the Imo River of Tanesuhara because large landslides that clog the river have occurred every century.

Table 2. Old landslides in Yamakoshi area.

Number	Year	Disaster	Place	Conditions
1	1824 April: thawing season	Landslide length 1.4 km, width 700 m	Nakano	Disaster throughout the village. 50% of rice fields could not be cultivated. Flood induced by river clogging. Two ponds of 100 m length and 10 m depth were created.
2	1926 May	Landslide length 1.1 km, width 180 m	Nakano	Fields and mountains of more than 2 km^2 were damaged. Floods induced by river clogging.
3	1929 April	Landslide width 50 m	Terano	Prefectural route (Tochio-Ojiya) was severed; bridge collapsed. River clogging is unclear.
4	1932	Landslide	Between Nakano and Terano	A large landslide occurred during construction of landslide measures. Details are unclear.

Geological Features

Related to unrecorded landslides of prehistoric times, the landslide history of Yamakoshi area was investigated using a topographical map that was compiled using information from an aerial photograph. Figure 4 portrays landslide locations during the Mid-Niigata Earthquake presented on the map of old landslide topography (provided by the National Research Institute for Earth Science and Disaster Prevention (NIED) and the Japan Science and Technology Agency (JST) [18]) in Yamakoshi area. Moreover, the figure was overlaid on a simplified geological map provided by Takeuchi et al. [19] and on landslide-designated areas obtained from a conservation map of Niigata prefecture [20]. Many designated landslide areas are located in the East Hill area. Large-scale landslides that occurred during the earthquake such as those of Terano or Higashi-takezawa might be of a reactivated type of landslide because they coincide completely with the old landslide topography [18].

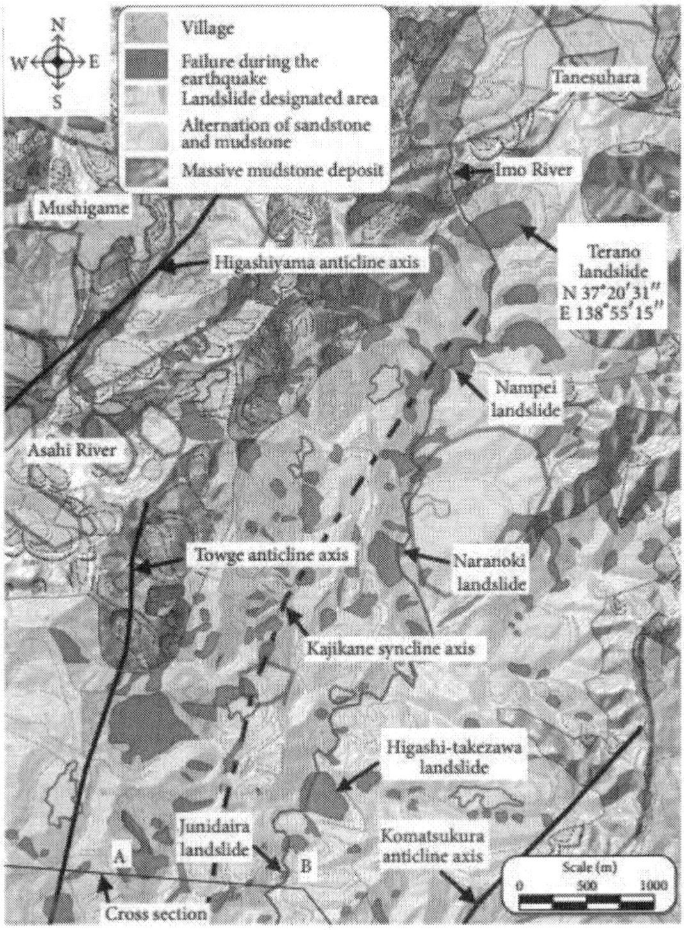

Figure 4. Landslide distribution and geological map around Yamakoshi area during Mid-Niigata Earthquake (provided by NIED and JST).

The west side of the map, which is classified in the Asahi River basin (Figure 4), is geologically an Araya deposit: massive dark grey mudstone. At the eastern side of the map, the Imo River basin (Figure 4), the deposit changes to alternating sandstone and massive mudstone, designated as Kawaguchi and Wanatsu deposits. Alternation of sandstone and mudstone is distributed mainly along the Imo River, except in its upper course. As presented in Figure 4, more numerous landslides occurred during earthquakes in areas with alternated layers of sandstone and mudstone than in areas with massive mudstone deposits. This finding implies that sandy natural slopes are more fragile than clayey natural slopes during earthquakes. However, landslide-designated areas, in which landslides are induced by snowmelt waters, are distributed mainly in the massive mudstone deposits. Moreover, the notable geological features of this region are syncline and anticline structures. They form a complex topography in which synclinal axes and anticlinal axes are arranged with a short interval (Figure 4). For that reason, peculiarly cuesta topography is apparent in this region. Fragile and weak slopes are therefore formed easily. In addition, the river scours the riverbed and the slope toe. Thereby, the slope becomes unstable. For those reasons, it is considered that numerous landslides occurred during the earthquake.

Figure 5 depicts a geological cross section of the area south of Yamakoshi area (Junidaira) published by the National Institute of Advanced Industrial Science and Technology (AIST). The location of this sampled geological profile is presented in Figure 4. Sandstone (W) and sandy mudstone (S, Ku_2) are distributed widely around the Imo River. This representative folded mountain area is composed of syncline and anticline. Geological cross sections clarify that a dip slope is apparent on the left bank of the Imo River. A reverse-dip slope is apparent on the right bank of the river. Peculiar slope failures occurred at the right and left banks of the Imo River. Surface failures occurred frequently at the right bank having reverse-dip slope strata, whereas problematical large landslides occurred occasionally at the left bank, which is a dip gentle slope.

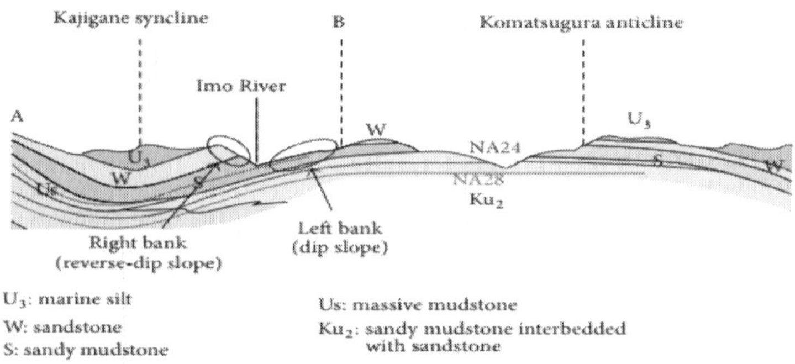

Figure 5. Geological cross section of the south of Yamakoshi area published by the National Institute of Advanced Industrial Science and Technology (AIST).

Mid-Niigata Offshore Earthquake

According to a report of the Ministry of Land, Infrastructure, Transport and Tourism (MLIT) given on 6 August 2007, 108 slope failures occurred during the Mid-Niigata Offshore Earthquake. The main slope failure sites are concentrated in the steep slope of coastal terrace from Shiiya to Hijirigahana, which is about 25 km distant, as shown in Figure 6. Although almost all slides were surface collapses, large mass movements rarely occurred in inland areas as described in Toyota and Onoue [13] because the earthquake occurred not under a mountainous area but under an offshore area. The fault extended from the epicentre to the south. However, the landslides are much fewer than the more than 3,000 which occurred during the Mid-Niigata Earthquake, despite the similarity of the earthquakes.

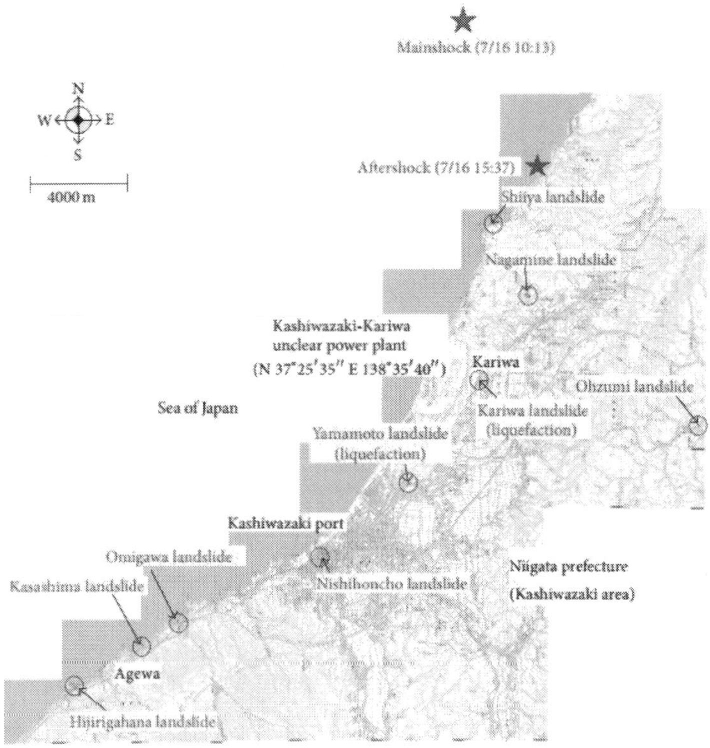

Figure 6. Landslide locations during the Mid-Niigata Offshore Earthquake (provided by GSI).

Northern Nagano Earthquake

The earth-slide disasters during the earthquake were not perfectly clear because the snow remained deeper than 2 m in April in mountainous areas. According to the Niigata prefectural office, 19 slope failures threatening residential life were reported on 20 March 2011. The slope failures were concentrated in Tokamachi city and Tsunan town, which have tertiary mudstone as their main geology (Figures 1 and 2). The Nagano prefectural office reported on 16 March 2011 that seven severe slope failures during the earthquake occurred in Sakae village, which has main geology of Andesitic rocks (Figures 1 and 2). Although the same type of earthquake occurred with the Mid-Niigata Earthquake that occurred directly underneath the mountainous area, it remains unclear why so few earth disasters were generated by the earthquake.

IMPORTANCE OF GROUNDWATER LEVEL

Water level and pore water pressure have been measured at the author's laboratory since 1995 in the old landslide area at the former Tochio city (Figure 1). The data were acquired using a data logger every hour on the hour. A detailed topographical map of this area is presented in Figure 7. The observation point is inferred to be the upper part of the landslide. The gentle slope of a river terrace, which was formed by the meandering flow of river, spreads in the lower part of the landslide. Many infiltration wells have been made in this area as landslide countermeasures. After their installation, no remarkable mass movement has been reported to date. Figure 8 shows the depth of ground-installed water pressure sensors and their geologic column. A borehole with a strainer, which is 27 m deep, was made in the slope. Then a water pressure transducer was set at 12 m depth in the borehole for estimation of water level. For pore water pressure measurement, a water-pressure transducer in the borehole was buried in sand. Their upper and lower sides were sealed with bentonite. The ground consists of soft silty soil up to 10.3 m deep from the surface and a sandy gravel layer of about 70 cm. Soft underlying rock comprises sand and silt in sections deeper than 11 m.

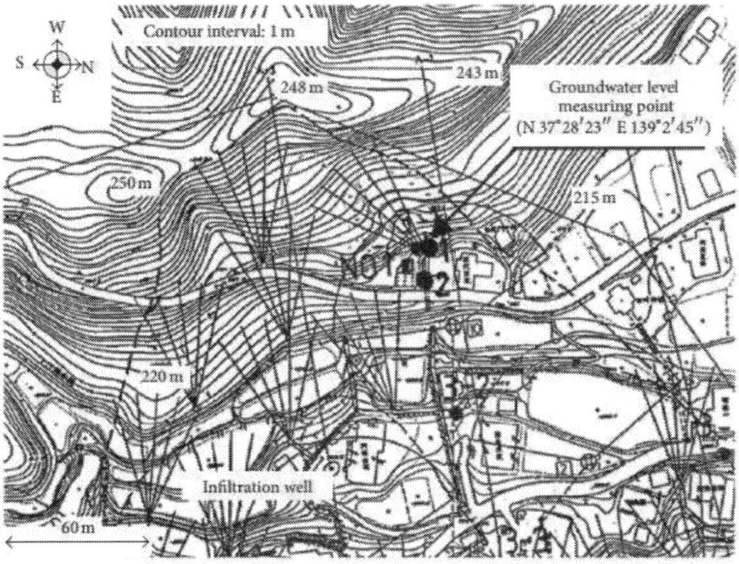

Figure 7. Topographical map (old landslide area at Tochio).

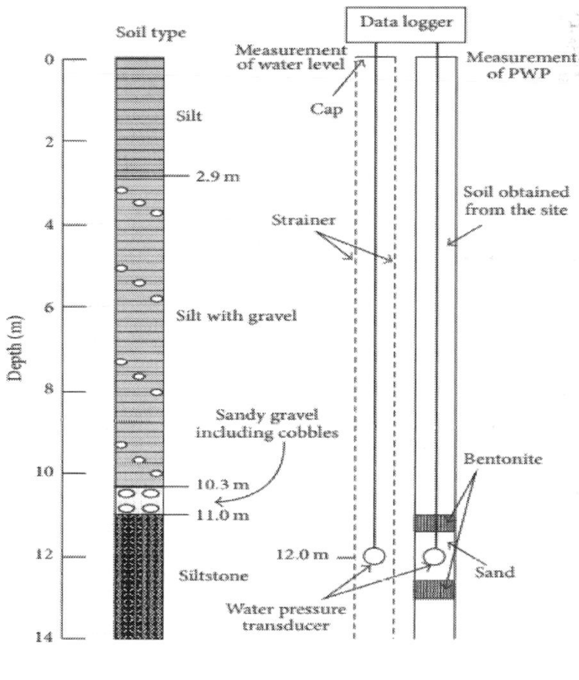

Figure 8. Depth of sensors and soil profiles.

Figure 9 presents records of groundwater level and hourly rainfall at Tochio during the Mid-Niigata Earthquake. Rainfall from typhoon number 23 occurred on 20 October 2004. The accumulated rainfall reached 100 mm in this area (Figure 9). The groundwater level rose quickly about 4 m with the rainfall. This area is inferred to include many catchment areas because the elevated groundwater levels descend gradually over a few days. The rise in the groundwater table induced by the typhoon had dropped by about 3 m in Tochio by the time of the Mid-Niigata Earthquake. It is apparent that this area still contained more water during the Mid-Niigata Earthquake than it did under ordinary conditions.

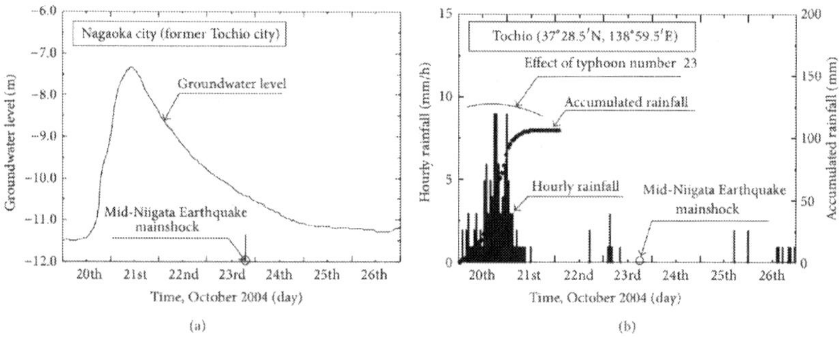

Figure 9. (a) Groundwater level and (b) rainfall during 20–26 October 2004.

Groundwater level fluctuation (October 2004–May 2005) is shown in Figure 10. The groundwater level at three days before the Mid-Niigata Earthquake was the highest in the displayed period of time. The snow season starts from mid-December. The total amount of snowfall reached 699 cm at Nagaoka during that winter. Snow melting accelerates from March with the coming of spring. The groundwater level increases during the snow-melting season because of soaking of snow water and suddenly drops after snow melting (in May), as shown in Figure 10. For that reason, tertiary type landslides have occurred frequently in this area in April and May: the snow-melting season. Therefore, it was feared that the slope damaged by the Mid-Niigata Earthquake starts moving again. Observations were continued during and after the snow season.

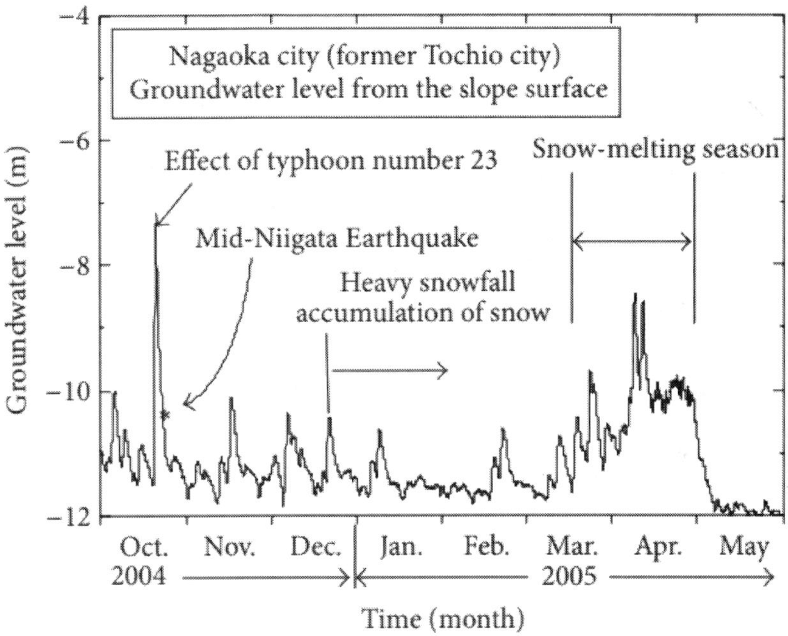

Figure 10. Groundwater level fluctuation.

FOLLOW-UP OF DAMAGED SLOPE

The Chuetsu region is a heavy snow area. The risk of snow avalanche has been increasing because the Mid-Niigata Earthquake damaged snow-protection facilities and natural vegetation. According to Niigata prefecture, 228 avalanche mitigation fences and 8 snow sheds were damaged during the earthquake. Figure11(a) shows an example of landslide in which soil mass slipped down with snow avalanche fences. Surface failures scoured the fence foundations. The fences were deformed because of falling soil. Soil accumulation behind fences and above snow sheds was also severe problem. As an emergency restoration, the accumulated soils were removed and large sandbags were emplaced to protect the infrastructure from the snow avalanche before the winter.

Figure 11. (a) Damage to snow avalanche fence; (b) snow-earth avalanche [21].

Snow avalanches occurred more frequently that year than in a normal winter. Some slopes were not covered with deep snow because a small snow slide occurred at all snowfalls. However, frequent patrols were conducted, preventing heavy damage. The risk of avalanche was reduced by the removal of unstable snow, construction of snow walls, and road closures during the winter. A new type of avalanche, where snow fell together with surface soil stripped from the damaged slope, was observed in Tokamachi City (Figure 1) as shown in Figure 11(b). An avalanche of this type, called a "snow-earth mixture avalanche," is regarded as a compound natural disaster, causing extensive damage to residences and roads [21].

Landslides have occurred frequently during the snow-melting season in this area because of high groundwater levels (Figure 10). Therefore, it was feared that the slope damaged by the Mid-Niigata Earthquake starts moving again as a secondary disaster. Therefore, after snow melting, follow-up investigations were conducted to observe the slope failure sites. However, Toyota [22] reported that no large mass movement occurred during the snow-melting season. These sliding masses induced by the Mid-Niigata Earthquake were therefore stable except for erosion from rain and snow melting. Situations of slope failures can be compared for those before winter and after the snow-melting season in Figure 12. There is no apparent difference between the photographs.

Figure 12. Slope failures: upper row, before winter, and lower row, after snow-melting season.

EARTHQUAKE DAMAGE COMPARISONS

Comparisons between the Mid-Niigata Earthquake, the Mid-Niigata Offshore Earthquake, and the Northern Nagano Earthquake are presented in Table 3. All earthquakes are mutually similar except for the frequency of the aftershocks: they were frequent after the Mid-Niigata Earthquake, but not after other earthquakes. Many evacuees, to escape from the fear of repeated strong aftershocks, were compelled to reside for extended periods in inconvenient evacuation areas. Some were afflicted with phlebothrombosis, often called "economy class syndrome."

Table 3. Comparisons of earthquakes.

	Mid-Niigata	Mid-Niigata Offshore	Northern Nagano
Date	23 Oct. 2004	16 July 2007	12 March 2011
Epicentre	Chuetsu area	Chuetsu offshore	Northern Nagano
Depth	13 km depth	17 km depth	8 km depth
Cause	Thrust fault	Thrust fault	Thrust fault
Magnitude	M6.8	M6.8	M6.7
JMA intensity	7 (at Kawaguchi)	6 upper (at Kashiwazaki)	6 upper (Sakae)
Max. acceleration	1750.2 gal (at Tokamachi)	812.7 gal (at Kashiwazaki)	803.5 gal (at Tsunan)
Aftershock	Frequent	Rare	Rare
Landslides	Thousands	Medium	Medium

However, the types of damage related to the earth disaster were different among the earthquakes in spite of being only 60 km distant from the epicentre. Although the main landslide sites are hilly and mountainous areas and although more than 3,000 slope failures occurred during the Mid-Niigata Earthquake, 108 slope failures occurred mainly in the steep slope of the coastal terrace during the Mid-Niigata Offshore Earthquake according to a report of MLIT on 6 August 2007 because the earthquake occurred not under a mountainous area but under an offshore area. The Northern Nagano Earthquake occurred under mountainous areas between Nagano and Niigata. However, few landslides were reported, as described in Section 3.3. Two possible reasons can be considered. The first reason is that the main geology of Sakae Village in the northern part of Nagano is Andesitic rock, which is different from the tertiary mudstone of the Chuetsu area (Figure 2). Examples of slope failures are exhibited in Figure 13. Figure 13(a) shows a river-clogging landslide at Yamakoshi area during the Mid-Niigata Earthquake. The weak soft deposit moved down and closed the Imo River channel. Figure 13(b) presents river-clogging landslides at Sakae Village during the Northern Nagano Earthquake. Apparently, the weathered part of a steep cliff fell and clogged a narrow valley. Then the dammed lake was created by the moving mass. However, the tertiary mudstone is distributed in Tokamachi and Tsunan areas (Figures 1 and 2) where strong seismic motion was recorded during the earthquake. Therefore, the second considered reason is that groundwater level might not be high during the Northern Nagano Earthquake compared with that of the Mid-Niigata Earthquake, as shown in Figure 9(a). From Figure 10, it is inferred that the ground water level remained low immediately before snow melting during the Northern Nagano Earthquake. Although snow remained deeper than 2 m in mountainous areas, few snow avalanches were triggered by the earthquake. Accumulated snow is apparently quite stable against earthquakes during this season (March). Consequently, the relations between the epicentre, the geography including geology, and groundwater level are extremely important factors to discuss to assess the risk of landslide damage.

Figure 13. Typical river-clogging landslides that occurred (a) during the Mid-Niigata Earthquake and (b) during the Northern Nagano Earthquake.

CONCLUSIONS

Several earthquakes struck the Chuetsu area of Niigata successively in recent years. Damage investigations have been reported individually and have been mutually compared. Information about damage related to geotechnical engineering has been assembled and discussed from the viewpoint of compound natural disasters. Although no severe damage has been certified as a secondary disaster, these disasters underscore the possibility of compound natural disasters or slight damage such as erosion occurring in the damaged slopes. A summary of the main findings from the study is presented below.(1)The importance of groundwater level during landslide disasters was assessed using field measurement results. Results show that groundwater during the Mid-Niigata Earthquake was higher than that under normal conditions.(2)Large-scale landslides during the Mid-Niigata Earthquake were regarded as reactivated type landslides because they occurred on old landslide traces.(3)Most landslides during the Mid-Niigata Earthquake occurred at alternating sandstone and mudstone strata, but landslides that moved gradually during the snow-melting season usually involved mudstone in large quantities from tertiary deposits.(4)Landslides during the Mid-Niigata Offshore Earthquake were fewer than those which occurred during the Mid-Niigata Earthquake, perhaps because the main tremor occurred not under a mountainous area but under an offshore area during the Mid-Niigata Offshore Earthquake.(5)Landslides that occurred during the Northern Nagano

Earthquake were also fewer than those that occurred during the Mid-Niigata Earthquake. The reasons considered are that the main geology of the mountainous area in northern Nagano is not tertiary mudstone but Andesitic rock. Moreover, the groundwater level might not be high compared to that prevailing during the Mid-Niigata Earthquake. Only a few snow avalanches triggered by the earthquake were reported in spite of snow that remained on the ground.(6)The risk of snow avalanche increased after the earthquakes because snow-protection facilities and natural vegetation were damaged.(7)Some additional damage occurred such as surface erosion and surface slides during rainfall and snow-melting season in the slopes damaged and loosened by the earthquakes. Carefully conducted daily observations are important to prevent a secondary disaster because some indications (slight damage) will precede severe damage. Timely countermeasures taken for cases of slight damage must be taken to reduce the risk of severe damage.

CONFLICT OF INTERESTS

The authors declare that there is no conflict of interests regarding the publication of this paper.

REFERENCES

1. Japan Society of Civil Engineers, Ed., Report on the 2004 Niigata Chuetsu Earthquake, CD Rom, Japan Society of Civil Engineers, 2006, (Japanese).
2. H. Toyota, J. Wang, K. Nakamura, and N. Sakai, "Evaluation of natural slope failures induced by the 2004 Niigata-ken Chuetsu Earthquake," Soils and Foundations, vol. 46, no. 6, pp. 727–738, 2006.
3. Y. Tsukamoto, K. Ishihara, and Y. Kobari, "Evaluation of run-out distances of slope failures during 2004 Niigata-ken Chuetsu Earthquake," Soils and Foundations, vol. 46, no. 6, pp. 713–725, 2006.
4. E. Rathje, R. Kayen, and K.-S. Woo, "Remote sensing observations of landslides and ground deformation from the 2004 Niigata Ken Chuetsu earthquake," Soils and Foundations, vol. 46, no. 6, pp. 831–842, 2006.
5. M. Chigira and H. Yagi, "Geological and geomorphological characetristics of landslides triggered by the 2004 mid Niigata prefecture earthquake in Japan," Engineering Geology, vol. 82, no. 4, pp. 202–221, 2006.

6. H. B. Wang, K. Sassa, and W. Y. Xu, "Analysis of a spatial distribution of landslides triggered by the 2004 Chuetsu earthquakes of Niigata Prefecture, Japan," Natural Hazards, vol. 41, no. 1, pp. 43–60, 2007.
7. A. Onoue, A. Wakai, K. Ugai et al., "Slope failures at Yokowatashi and Nagaoka College of Technology due to the 2004 Niigata-ken Chuetsu Earthquake and their analytical considerations," Soils and Foundations, vol. 46, no. 6, pp. 751–764, 2006.
8. J. Deng, H. Kameya, Y. Miyashita, J. Kuwano, R. Kuwano, and J. Koseki, "Study on dip slope failure at Higashi Takezawa induced by 2004 Niigata-Ken Chuetsu earthquake," Soils and Foundations, vol. 51, no. 5, pp. 929–943, 2011.
9. K. Sassa, H. Fukuoka, G. Scarascia-Mugnozza, and S. Evans, "Earthquake-induced landslides: distribution, motion and mechanisms," Soils and Foundations, pp. 53–64, 1996.
10. N. M. Newmark, "Effects of earthquakes on dams and embankments," Géotechnique, vol. 15, no. 2, pp. 139–160, 1965.
11. Japanese Geotechnical Society, Ed., Report on the 2007 Niigata-ken Chuetsu-oki Earthquake, Japanese Geotechnical Society, 2009, (Japanese).
12. A. Onoue and H. Toyota, "Damage induced by the 2007 Niigataken Chuetsu-oki earthquake," inProceedings of the 14th World Conference on Earthquake Engineering, Paper ID: S26-15, pp. 1–8, Beijing, China, 2008.
13. H. Toyota and A. Onoue, "Characterisation of slope failures during the 2004 Niigata-ken Chuetsu and the 2007 Niigata-ken Chuetsu-oki Earthquake," in Proceedings of the 14th World Conference on Earthquake Engineering, PaperID: S26-18, pp. 1–8, Beijing, China, 2007.
14. Joint Editorial Committee for the Report on the Great East Japan Earthquake Disaster, Report on the Great East Japan Earthquake Disaster, Fundamental Aspects 3, Geohazards (Abstract), JGS, 2014, (Japanese).
15. Japan Metrological Agency, "Report on the Northern Nagano Earthquake (2nd issue)," 2011,http://www.jma.go.jp/jma/press/1103/12d/201103120800.html.
16. Niigata Branch of the Japan Landslide Society, Ed., Record of Landslide Disasters in Niigata, CD Rom, Niigata Branch of the Japan Landslide Society, 2003, (Japanese).
17. Editorial Committee of History of Yamakoshi Village, History of Yamakoshi Village, Editorial Committee of History of Yamakoshi Village, Yamakoshi, Japan, 1981, (Japanese).
18. National Research Institute for Earth Sciencc and Disaster Prevention (NIED) and Japan Science and Technology Agency (JST), "Landslide map," 2004,http://lsweb1.ess.bosai.go.jp/jisuberi/jisuberi_mini/jisuberi_top.html.

19. K. Takeuchi, Y. Yanagisawa, J. Miyazaki, and M. Ozaki, "1:50,000 digital geological map of the Uonuma region, Niigata Prefecture (Ver. 1)," GSJ Open-File Report 412, 2004,http://www.gsj.jp/GDB/openfile/files/no0412/index.html.
20. Niigata Prefecture, Ed., Conservation Map, Niigata Prefecture, Niigata, Japan, 1982, (Japanese).
21. I. Kamiishi and M. Machida, "Snow avalanche by the Chuetsu Earthquake and heavy snow," inEarthquake Disaster and Prevention in Snowy Areas, pp. 45–67, Japan Society for Snow Engineering, Joshinetsu, Japan, 2008, (Japanese).
22. H. Toyota, "Stability progress of slopes damaged by natural disaster—case study in the Chuetsu area of Niigata, Japan," in Proceedings of the 17th Southeast Asian Geotechnical Conference, vol. 2, pp. 113–122, Taipei, Taiwan, 2010.

CITATION

Hirofumi Toyota and Susumu Takada, "Geotechnical Distinction of Landslides Induced by Near-Field Earthquakes in Niigata, Japan," Geography Journal, vol. 2015, Article ID 359047, 11 pages, 2015. doi:10.1155/2015/359047

CHAPTER 6

Optimization Design of Foundation Excavation for Xiluodu Super-High Arch Dam in China

Qixiang Fan, Shaowu Zhou, , Ning Yang

China Three Gorges Corporation, Beijing 100038, China

ABSTRACT

With better understanding of the quality and physico-mechanical properties of rocks of dam foundation, and the physico-mechanical properties and structure design of arch dam in association with the foundation excavation of Xiluodu arch dam, the excavation optimization design was proposed for the foundation surface on the basis of feasibility study. Common analysis and numerical analysis results demonstrated the feasibility of using the weakly weathered rocks III_1 and III_2 as the foundation surface of super-high arch dam. In view of changes in the geological conditions at the dam foundation along the riverbed direction, the design of extending foundation surface excavation area and using consolidating grouting and optimizing structure of dam bottom was introduced, allowing for harmonization of the arch dam and foundation. Three-dimensional (3D) geomechanics model test and finite element analysis results indicated that the dam body and foundation have good overload stability and high bearing capacity. The monitoring data showed that the behaviors of dam and foundation correspond with the designed patterns in the construction period and the initial operation period.

INTRODUCTION

As is known, arch dam basically receives forces through both the dam body and the foundation. As the foundation of an arch dam, the rock mass has to bear not only the deadweight of the upper arch dam, but also the external loads such as water pressures and silt loads, ensuring the stability of itself and the safety of the whole arch dam (Li et al., 2004 and Wang et al., 2010). A super-high arch dam over 200 m bears significantly large external loads, for which a matching dam foundation is desirable. For example, hydraulic thrust of Xiluodu super-high arc dam is estimated to be more than 140×10^6 kN (HCEC, 2005). However, it is always very difficult to accurately evaluate the complicated hydrogeological and other conditions of a natural dam site due to the limitations in geological survey, especially the sites on the riverbed and valley floor in a gorge area. In a common sense, the foundation surface excavation is a process of dynamic adjustment, which should adapt to the dynamic adjustment of the dam body shape. For an arch dam over 200 m, the foundation surface excavation will be generally 6–12 months delayed from onset of a project. Therefore, the optimization design for foundation excavation of super-high arch dam is critically important.

In China, the rock mass at a dam site can be roughly classified as completely weathered zone, intensively weathered zone, and weakly weathered zone and slightly weathered to fresh zone, from the outside to the inside. Obviously, the first two categories cannot be used as the rock foundation of a high arch dam, and thus must be excavated. The weakly weathered to fresh rocks can be used directly as the foundation surface of a high arch dam. Generally, the deeper foundation means higher integrality of the foundation rocks, thus it can bear larger capacity in terms of global safety. However, the foundation excavation and cast-in-place concrete amount will be significantly increased, leading to a much longer construction period and larger additional investment. On the other hand, the external loads on the dam body and dam abutment will increase with increasing depth of dam abutment and arch span, and those problems such as high slope stability will arise. Therefore, the process of optimization design for the foundation of a super-high arch dam is focusing on the feasibility and utilization degree of weakly weathered rock mass as dam foundation with increasing understanding of the geological conditions, and thus to establish a suitable evaluation system for safety purpose of arch dam body and foundation (Wang, 2007 and Wang and Lin, 2008).

This paper introduces the optimization concepts of the foundation interface design of the Xiluodu super-high arch dam, the structure optimization design and foundation treatment measures in terms of variations in basic geological conditions in the riverbed region. The finite element analyses and three-dimensional (3D) geological model test results indicate that when using the weakly weathered rocks III_1 and III_2 as the foundation surface of the riverbed dam, the structure optimization at the bottom of the riverbed dam section and the foundation treatment measures adopted can meet the requirements of the overall stability of the dam and foundation, which is also verified by the monitoring data recorded in the construction period and the initial impoundment and operation periods.

UNDERSTANDING AND OPTIMIZATION PRINCIPLES OF SUPER-HIGH ARCH DAM FOUNDATION SURFACE

Understanding of super-high arch dam foundation surface

According to China's standards (SL282-2003, 2003 and DL/T5346-2006, 2007), the base rocks in the fresh to slightly weathered zone or the lower part of the weakly weathered zone can be used as the foundation surface for an arch dam, and a special study must be conducted when necessary. In the arch dam design, the degree of rock mass weathering is always used to determine the depth of dam foundation excavation, without concrete control standards for the modulus of deformation. While in other countries and/or regions, more emphases are focused on the effect of foundation reinforcement on the dam body for the technical and economic purposes. Only when the reinforcement measures fail to achieve the desired effect, the entire rock area above the foundation surface will be removed (DIBR, 1977, SNiP 2.06.06-85, 1985 and Zhu, 1988). Some arch dams are built on foundation with very low deformation modulus. For example, the Vajont arch dam is 262 m high, with a foundation deformation modulus of 7.5–16 GPa. Another case is the 237 m high Mauvoisin dam, which has a foundation deformation modulus of 4 GPa approximately.

In China, as the understanding of high arch dam is increased and various engineering experiences are obtained, more and more high arch dams begin to use weakly weathered rocks (III_1 or III_2) as foundation surface, and the desirable foundation reinforcement measures are employed to meet the requirement of safe engineering operation, such as the Ertan arch dam, of which the indicators of foundation surface rocks are shown in Table 1 (CHIDI, 2000). This is also reflected in the revised arch dam design standards (SL282-2003, 2003 and DL/T5346-2006, 2007).

Table 1. Rock mass classification and basic characteristics of Ertan arch dam foundation.

Rock classification	Rock lithology	Foundation deformation modulus (10^4 MPa)	Rock percentage (%)	Wave velocity V_P (10^4 m/s)
A	Syenite	3.5	24	0.6
B	Basalt	1–3.5		
C	Syenite/Basalt	1–1.5	57	0.5
D	–	0.5–0.8	18	0.3–0.45
Weak belt of fault			1	

Optimization principles and steps of super-high arch dam foundation surface

For the design of general high arch dam, using weakly weathered rocks for some parts of the dam foundation surface is generally acceptable, which can be verified by the common stress analysis method, calibrated by foundation anti-sliding stability analysis and foundation reinforcement methods (Wang, 2007 and Wang and Lin, 2008). But for super-high arch dams built in remote mountainous and gorge areas and/or high seismic intensity areas, the geological survey for the foundation of riverbed dam section may be insufficient in the feasibility study stage in comparison to the technical construction stage. Due to the possible drawbacks in geological survey and technical measures, insufficient analysis will be induced, resulting in an unreasonable design. At the same time, the weakly weathered rocks may have a certain degree of effect on the arch dam stability and safety, due to the slightly poor properties and weaker physico-mechanical behaviors. Thus the rock mass properties, structural safety of arch dam, anti-sliding stability, and overall arch dam safety should be combined to evaluate how to utilize the weakly weathered rocks, especially for the construction of a 300 m super-high arch dam.

Referring to the foundation excavation process of Xiluodu arch dam, it is suggested that the feasibility is desirable that how to use the rocks III_1 and III_2 as the foundation surface in order to optimize the design of super-high arch dam foundation, for the purposes of safety and reliability. Detailed steps are described as follows:

1) Collect data of site-specific hydrogeological conditions in the dam site, evaluate the quality of rock mass, and determine the rock quality grading and mechanical parameters of rock mass.

2) Conduct a preliminarily study on the foundation treatment measures, evaluate the effect of consolidation grouting on the physico-mechanical properties of rock mass, and select the values of foundation rock mass parameters, such as foundation deformation modulus after consolidating grouting is used.
3) Use the engineering analogy method to analyze the feasibility of foundation optimization, and propose 2–3 candidate foundation surfaces for optimization design of arch dam.
4) Preliminarily evaluate the feasibility of foundation optimization through commonly used safety analysis methods, such as static stress and dynamic stress method, arch-cantilever method (ACM), stress–displacement analysis method, and anti-sliding stability analysis method.
5) Evaluate the stability of dam body and foundation after optimization of foundation surface through numerical analysis (e.g. finite element method) and 3D geomechanics model test, and further verify the reasonability of optimization excavation of foundation surface.
6) Determine the optimization design scheme in terms of comprehensive analysis and comparison.
7) Qualitatively evaluate the quality of foundation rock surface at the excavation stage; at the same time, optimize and adjust the foundation treatment measures, with special focus on the geological defects, inter-laminar and intra-formational dislocation belt, according to the site-specific geological conditions.
8) Optimize the design of dam structure at the riverbed dam section, and increase its rigidity adapted to the foundation deformation; use the 3D nonlinear finite element method or geomechanics model test to evaluate the stability and safety of dam foundation with optimized arch dam structure.
9) Verify the reasonability of excavation optimization method of super-high arch dam foundation by using monitoring data and the feedback result of arch dam in the construction and (initial) operation periods.

OPTIMIZATION DESIGN OF FOUNDATION EXCAVATION OF XILUODU ARCH DAM

Analysis of the influential factors on foundation surface selection

The watercourse of Xiluodu dam site is characterized by huge mountains, abrupt slopes, and complete landform on both banks, but with no deep trenches. The valley cross-section is in a narrow U-shape, and the valley width ratio is about 2.0. The angle of the left-bank valley slope is 40°–75°, and the right-bank valley slope is 55°–75°, steep in the upper part and gentle in the lower part. The foundation surface selection is not governed by the landform, and Fig. 1 shows the topography of Xiluodu dam site.

Figure. 1. Topography of Xiluodu dam site.

Base rocks of the dam site are composed of Emeishan basalt of the Upper Permian System, characterized by hard rock mass, high strength, massive structure, and complete to relatively complete blocks. Rock mass in the dam area is classified into five categories and seven sub-categories in terms of comprehensive quality. Table 2 lists the physico-mechanical parameters. The structural surface of weak steep angles controls the dam foundation deformation and/or sliding of dam abutment anti-force body, which is not developed with hard to relatively hard rock mass in terms of sufficient strength. The rocks IV and V with poor integrity and cataclastic structure are weakly weathered unloaded rock mass, which are not suitable for the target foundation. Whilst the rocks I and II characterized with sufficient strength, and rock mass III_1 characterized with deformation modulus of 10–12 GPa and mediate anti-deformation capacity, can be used as the dam foundation. Rocks III_2 at some sections can also be used after being reinforced.

Table 2. Comprehensive rock mass quality classification and mechanical parameters in the dam area.

Rocks	Sub-categories	Rock mass structure type	Rock mass weathering	Degree of development L_c	Wave velocity V_P (m/s)	Deformation modulus E_0 (GPa)		Elasticity modulus E (GPa)	
						Horizontal	Vertical	Horizontal	Vertical
I		Massive	Fresh	Not developed	>5500	24–36	24–36	33–50	33–50
II		Massive	Slightly weathered to fresh	Weakly developed	4800–5500	17–26	12–16	22–30	16–22
III	III_1	Massive to sub-massive	Lower weakly weathered section, partially slightly weathered to fresh	Weakly developed–relatively developed	4000–5200	9–16	9–11	14–20	13–16
	III_2	Mosaic	Upper and lower weakly weathered sections, partially slightly weathered to fresh	Relatively developed–developed	3500–4500	5–7	4–6	7–9	5–8
IV	IV_1	Mosaic–cataclastic	Upper weakly weathered section, partially lower weakly weathered section		2700–4000	3–4	2.5–3.5	4–5	4–5
	IV_2	Cataclastic	Upper weakly weathered section, partially slightly weathered to fresh	Developed	2500–4000	0.9–2	0.5–1	1–2.6	0.7–1.2
V		Bulk	Intensively weathered		<2500	0.5–0.8	0.3–0.4	0.7–1.1	0.4–0.5

OPTIMIZATION DESIGN OF FOUNDATION BASED FEASIBILITY STUDY STAGE EXCAVATION

Optimization design of foundation excavation

The selection of arch dam foundation and determination of excavation depth have major effects on dam safety, work amount, construction period and investment (Liu, 1996). At the feasibility study stage, the riverbed dam foundation of Xiluodu arch dam was placed at the elevation of 332 m, dominated by weakly weathered to fresh rock mass II. The weakly weathered rock mass III_1 of lower section was only used when reaching the upper elevation. With further understanding rock quality and mechanical properties of the arch dam foundation associated with the mechanical properties and structure design of arch dam, the design of arch dam foundation was optimized by referring to the established engineering experiences, in order to minimize the arch abutment depth and use the rocks III_1 and III_2 on the premise of the safety of arch dam foundation.

(1) Riverbed foundation

According to the supplementary results of geological survey, the rock mass on riverbed is considered to be good quality globally, with a small amount of treatment work for dam foundation (see Fig. 2). After detailed analysis, the elevation of 332 m was used as the arch dam riverbed foundation in the optimization and comparative schemes, which was raised to 340 m at some places on both sides. Partial intra-formational dislocation belts under the foundation surface and developed weak inter-layers were reinforced; and partial rock mass was removed and concrete was backfilled at some places. Table 3 lists the horizontal depth of the downstream arch abutment of the riverbed foundation for three kinds of foundation surfaces.

(2) Foundations on both banks

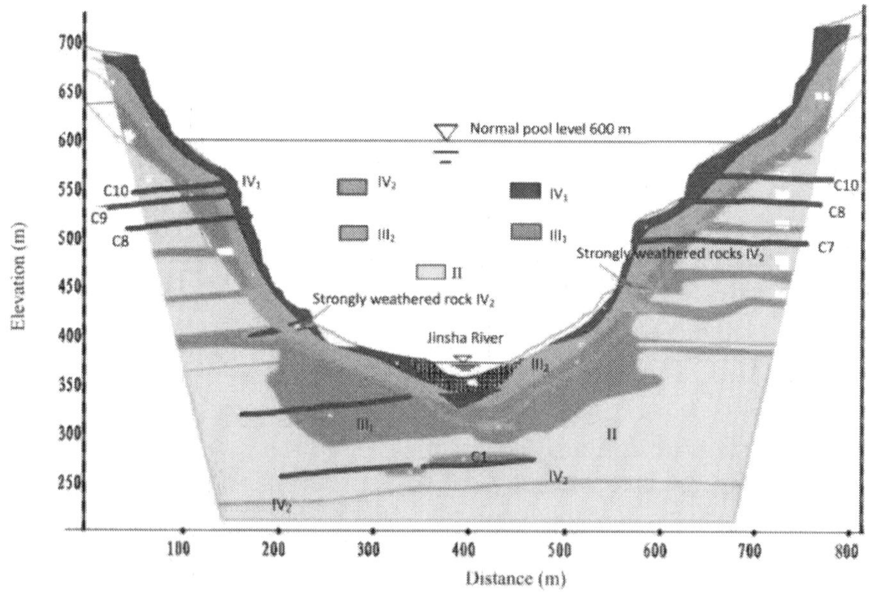

Figure. 2. Engineering geological profile of the Xiluodu arc dam.

Table 3. Horizontal depth of downstream arch abutment of riverbed foundation arch support (unit: m).

	Feasibility study scheme		Optimization scheme		Comparative scheme	
Elevation	Left bank	Right bank	Left bank	Right bank	Left bank	Right bank
370	83.3	98.3	64.4	55.6	62.8	54.5
350	62.7	77.1	44.2	36.5	40.2	35
Average	73	87.7	54.3	46.1	51.5	44.8

For the feasibility study scheme, the dam foundation mainly was composed of the slightly weathered to fresh rocks II; whilst weakly weathered rock mass III_1 was used only at some places at the elevation of over 560 m.

For the optimization scheme, the dam foundation was moved slightly outside to reduce the depth of arch abutment, with weakly weathered rock mass III_1 as foundation surface, and rock mass III_2 as foundation surface for some places at the elevation of over 560 m on the left and right banks.

For the comparative scheme, on the basis of the optimization scheme, the foundation was further adjusted slightly to reduce the depth, and increase the exposure range of rock mass III_2. Fig. 3 displays the depth of arch abutment of three kinds of foundation surfaces, and Table 4 lists the rock mass quality classification.

(3) Comprehensive comparative analysis of various schemes

i. Rock formation analysis

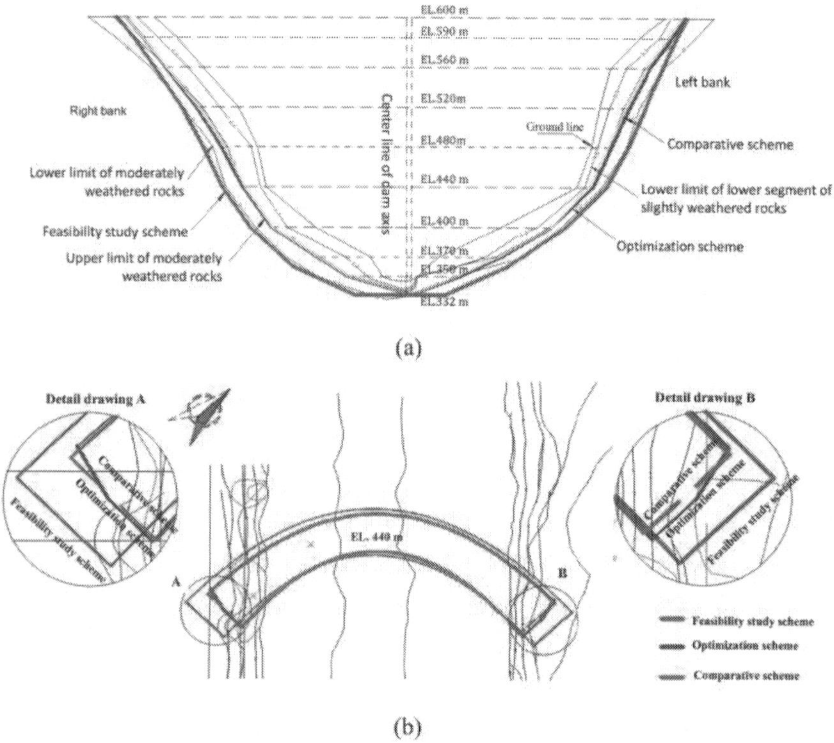

Figure. 3. Arch abutment depth of three kinds of foundation surfaces. (a) Downstream vertical view of arch abutment depth for three kinds of foundation surfaces. (b) Horizontal sliced view of arch ring for three kinds of foundation surfaces at elevation of 440 m.

Table 4. Dam foundation rock mass quality classification of 3 kinds of foundation surfaces.

	Percentage of different rock categories (%)				
Scheme	I	II	III_1	III_2	IV_1
Feasibility study scheme	–	69.63	30.29	0.08	–
Optimization scheme	–	23.13	61.89	12.3	2.67
Comparative scheme	–	21.22	59.35	15.3	4.13

Feasibility study scheme: on the foundation surface, the area of rocks II accounted for more than 69%, rocks III_1 accounted for 30% approximately; rocks III_2 were only exposed at the downstream side at the elevation of 610 m of the left bank. From the downstream vertical view of the arch abutment, except for the elevation of over 520 m and elevation of 400 m where weakly weathered rock mass III_1 was used, the arch abutment was basically placed on the slightly weathered to fresh rock mass II on the left bank.

Comparative scheme: the area of rock mass II accounted for 21.22%, rock class III_1 for 59.35%, rock mass III_2 for 15.3%; rock mass IV_1 was only exposed at some places at the elevation of 400–410 m on the left bank and near the elevation of 420 m on the right bank, accounting for 4.13%.

Optimization scheme: compared with the comparative scheme, around 2% rocks II and III_1 were increased, rock mass III_2 decreased roughly by 3%, rock mass IV_1 decreased by 1.5%. The quality of arch dam foundation rock mass in the optimization scheme was better than that in the comparative scheme, and slightly worse than that in the feasibility study scheme (see Table 4 and Fig. 4).

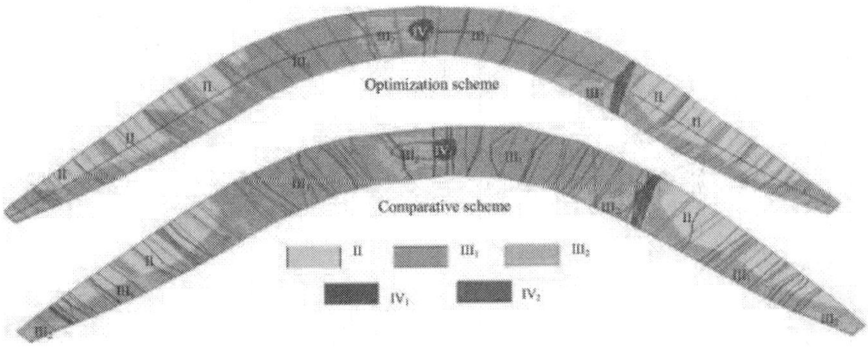

Figure 4. Foundation rock mass quality classification in optimization and comparative schemes.

(ii) Analysis of foundation depth

Feasibility study scheme: the depth of arch dam foundation on the left and right banks was relatively large, with an average of 43.6 m and 49.7 m, respectively, but favorable formation conditions of Xiluodu dam site were not fully used.

Comparative scheme: compared with the feasibility study scheme, the average depth was significantly decreased, about 16 m and 14 m on the left and the right banks, respectively, especially at the lower middle elevation. The arch abutment below the elevation of 560 m was basically placed on the lower segment weakly weathered rocks III_1, and upper segment weakly weathered rocks III_2 was used at the elevation over 560 m. The foundation surface shape was relatively smooth.

Optimization scheme: the left-bank arch abutment depth was about 3 m deeper than that in the comparative scheme. The right-bank one was basically equal to that in the comparative scheme. For the abutment of the left-bank arch at elevation of 440 m, it was 4.4 m deeper than that in the comparative scheme, thus more rock mass III_1 and less rock mass III_2 can be used in order to control the dam foundation deformation associated with enough bearing capacity. The right-bank arch abutment depth within the middle elevation of 560–440 m was increased slightly (see Table 5).

Table 5. Comparison of horizontal depth of downstream arch abutment for 3 kinds of foundation surfaces.

Scheme	Bank	Horizontal depth at various elevations (m)							
		EL 610 m	EL 590 m	EL 560 m	EL 520 m	EL 480 m	EL 440 m	EL 400 m	Average
Feasibility study scheme	Left bank	22.2	39.6	54.1	47.2	41.4	30.4	70.1	43.6
Optimization scheme		20.4	34.4	40.8	30.1	21.4	16.4	51.3	30.7
Comparative scheme		17.2	31.4	37.9	27.3	16.8	12	50	27.5
Feasibility study scheme	Right bank	31.4	45.2	49.1	56.8	57.2	45.2	63.2	49.7
Optimization scheme		29.3	36	38.4	43	37.5	24.6	45.7	36.4
Comparative scheme		29.4	36	37.2	41.1	34.4	23.4	45	35.2

Design of arch dam shape

The ACM was used to calculate the dam stress. Five line shapes (parabola, unified quadratic curve, 3-center circle, log-spiral, and ellipse) were compared and analyzed in consideration of engineering geology, design, construction, engineering experience and other factors. It is reported that parabola was chosen as the line type of Xiluodu arch dam as it is better adaptable to U-shaped valley (HCEC, 2005). Table 6 gives the dam characteristic parameters for three schemes.

Table 6. Parameters of parabola-shaped double-curvature arch dam.

Scheme	Thickness of arch crown top (m)	Thickness of arch crown bottom (m)	Max. thickness of arch abutment (m)	Arc length of vault center line (m)	Max. central angle (°)	Thickness ratio	Arc height ratio	Upstream overhang degree	Flexibility coefficient	Concrete volume (10^4 m^3)	Dam foundation excavation volume (10^4 m^3)
Feasibility study scheme	14	69	75.7	698.07	96.21	0.248	2.512	0.217	10.68	665.6	526
Optimization scheme	14	60	64	681.49	95.13	0.216	2.451	0.141	11.1	555.29	365
Comparative scheme	14	60	64	678.65	95.36	0.216	2.441	0.125	11.29	501.74	314.7

Compared with the feasibility study scheme, the arc length of arch dam vault center line and the arch dam thickness in the optimization scheme (see Fig. 5) were significantly decreased, but the arc length of center line and the upstream overhang degree were increased slightly, and other shape parameters were basically the same. Partial adjustment of the arch abutment position of foundation almost had no effect on the design of arch dam shape. The dam foundation excavation was 1.61 million m^3 less and the dam concrete amount was 1.10 million m^3 less in the optimization scheme than those in the feasibility study scheme.

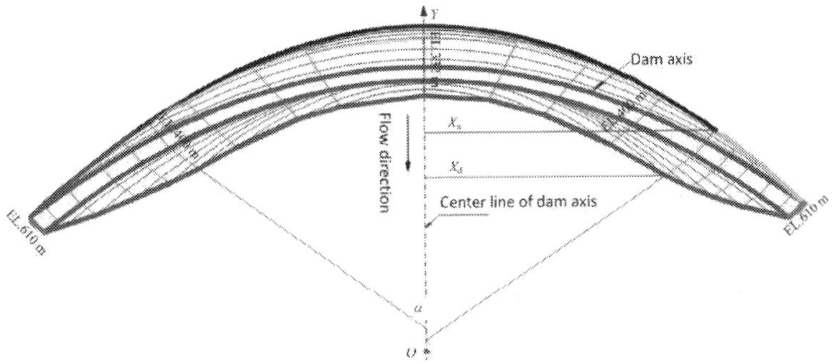

Figure. 5. Design of arch dam shape in optimization scheme.

Safety evaluation and analysis

With the basic load combinations, the calculated results indicate similar stress distribution of dam in the three schemes. The dam face was basically in a condition of compression state, meeting the stress control standards, i.e. the maximum tensile stress should be not larger than 1.2 MPa, the allowable compressive stress is not larger than 9.0 MPa, and safety coefficient of concrete strength under compression is 4.0. In the optimization scheme, the maximum compressive stress of the upstream dam surface was 7.56 MPa, and that of the downstream dam face is 8.66 MPa. The maximum principal tensile stress of the upstream dam face was 1.13 MPa, and that of the downstream dam face is 1.04 MPa. At the same time, analysis of deformation modulus sensitivity of host rocks (±20% deformation modulus) shows that the changes in the foundation deformation modulus had no significant effect on the dam stress, and the foundation adaptability was good in each scheme.

Three kinds of foundation surfaces show a slight difference in the analysis results of anti-sliding stability of dam abutment. Both the factors of safety under pure friction and shear friction conditions basically meet the design control standards. Table 7 presents the factors of safety of anti-sliding dam abutment in the optimization scheme.

Table 7. Factors of safety of anti-sliding dam abutment in the optimization scheme.

Bank	Mass shape	Pure friction				Shear friction			
		Partial failure of curtain drainage		Normal curtain drainage		Partial failure of curtain drainage		Normal curtain drainage	
		Min.	Max.	Min.	Max.	Min.	Max.	Min.	Max.
Left	Bulk mass	1.39	2.43	1.44	2.48	3.76	9.64	3.83	9.71
	Stepped mass	1.45	1.52	1.52	1.58	3.72	4.04	3.8	4.11
Right	Bulk mass	1.6	-	1.65	-	3.42	-	3.5	-
	Stepped mass	2.06	2.31	2.15	2.39	4.39	5.11	4.49	5.2
Standard		≥1.3				≥3.5			

3D nonlinear finite element analysis and 3D geomechanics model test results (Zhou et al., 2004 and Zhou et al., 2006) show that the foundation optimization will not cause abnormality of the dam stress variation. The arch dam stress (see Table 8), displacement distribution, overall safety for 3 kinds of foundation surfaces were quite promising. Meanwhile the overload analysis shows that the global dam structure under normal load combinations was in a state of elastic working condition. No yield and cracking damage appeared on both the upstream and downstream dam faces. This was mostly benefitted from the symmetrical U-shaped valley of Xiluodu dam site. Table 9 presents the relevant crack initiation overload coefficient K_1, nonlinear deformation load coefficient K_2, and ultimate overload coefficient before overall instability K_3 of 3 kinds of foundation surfaces. It can be seen that the dam overload capacity in the optimization scheme was equal to that in the feasibility study scheme, and better than that in the comparative scheme, indicating a strong relation between the overload capacity and the quality of foundation rock mass.

Table 8. Arch dam stress distribution value for 3 kinds of foundation surfaces (3D nonlinear finite element).

Scheme	Upstream face				Downstream face			
	Maximum tensile stress of dam heel (MPa)	Maximum pulling stress along dam axis (MPa)	Maximum tensile stress of left arch abutment (MPa)	Maximum tensile stress of right arch abutment (MPa)	Maximum compressive stress of dam toe (MPa)	Maximum compressive stress of left arch abutment (MPa)	Maximum compressive stress of right arch abutment (MPa)	Maximum compressive stress of dam face (MPa)
Feasibility study	0	−6.65 (EL 500 m)	0.91 (EL 500 m)	1.03 (EL 580 m)	−13.3	−16.3 (EL 380 m)	−15.2 (EL 360 m)	0.75 (EL 560 m)
Optimization	0.95	−7.54 (EL 500 m)	0.96 (EL 560 m)	0.78 (EL 500 m)	−12.8	−18 (EL 380 m)	−17.5 (EL 360 m)	0.66 (EL 480 m)
Comparative	0	−7.54 (EL 480 m)	1.05 (EL 500 m)	1.12 (EL 500 m)	−10.2	−18.9 (EL 360 m)	−19.8 (EL 360 m)	0.16 (EL 560 m)

Table 9. Arch dam foundation overload coefficient under 3 kinds of foundation surfaces.

Analysis method	Scheme	K_1	K_2	K_3
3D nonlinear finite element	Feasibility study	2	4	8
	Optimization	2	3–4	7.5
	Comparative	1.8–2	2.5–3	5
3D geomechanics model test	Feasibility study	1.8	4–5	6.5–8
	Optimization	1.8–2	4.5	8.5

In general, 3 kinds of foundation surfaces could all meet the requirements of arch dam design for foundation surface, indicating a good adaptive capacity of arch dam to the changes in the dam foundation deformation modulus. In terms of overload capacity, both the feasibility and optimization schemes can meet the design requirements, but the comparative scheme is slightly defective. The optimization scheme could keep the elevation of dam foundation on the riverbed unchanged, with two banks being moved outward appropriately for engineering safety and significant economic benefits purposes.

Optimization design of dam riverbed foundation structure

Global enlargement of excavation of riverbed dam foundation

As the excavation continued, replacement excavation was conducted for geological defects (8 areas on the left bank and 9 areas on the right bank) of the inter-laminar and intra-formation dislocation band and intensively weathered inter-layers that were developed in the basalt rocks at the elevations of 610–500 m, 500–400 m and below 400 m, as well as the weakly weathered rocks III_2 and weak unloaded rocks IV exposed to the foundation surface (Wang et al., 2007a, Wang et al., 2007b, Wang et al., 2007c,Wang et al., 2008, Yang et al., 2007a and Yang et al., 2007b). Fig. 6 and Fig. 7 illustrate the overall arrangement and cross-section of foundation replacement, respectively. After replacement excavation, the foundation at the elevation of 560–610 m was dominated by rocks II and III_1, some by rocks III_2. The abrupt-slope dam section foundation at the elevation of 560–400 m was dominated by rocks II and III_1. Rocks III_2 accounted for 5.3% on the left bank, and only 0.8% on the right bank; whilst rocks III_1 accounted for more than 90% on the gentle-slope dam section foundation at the elevation of 400–328 m, meeting the requirements of foundation rock mass quality.

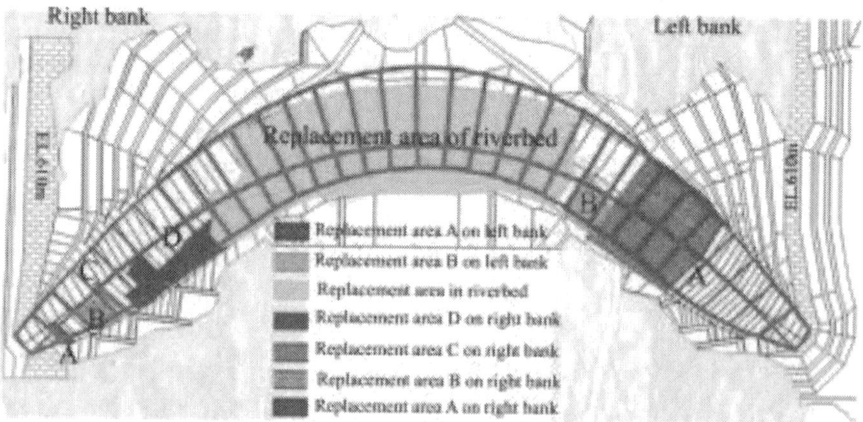

Figure. 6. Overall arrangement of foundation replacement.

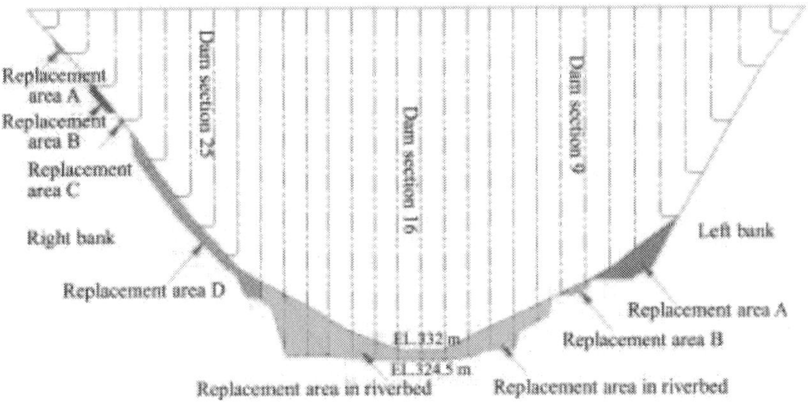

Figure. 7. Typical cross-sectional view of foundation replacement.

Additional geological survey of riverbed dam section reveals that the rocks were strongly weathered and the structure of rock mass was not uniformly distributed in the riverbed sections and the gentle-slope dam sections on both banks below the elevation of 400 m, where the disturbed belt was developed in association with gentle slope attitude, especially in the inter-layer along the dislocation belt. Despite that rocks III_1 were of a large portion, rocks III_2 and weathered inter-layers in the zone were widely distributed. Given that rock mass in the riverbed dam section is the key of foundation deformation, the riverbed dam foundation had to be raised from the elevation of 340–332 m to 328–324.5 m through one-step global enlargement of excavation, in order to ensure the long-term safe operation of the arch dam. Therefore, the number of the riverbed dam sections of foundation was increased from 3 (dam sections 15–17) to 6 (dam sections 14–19).Fig. 8 shows the overall view of foundation surface of Xiluodu arch dam and Fig. 9 shows the overall view of riverbed foundation surface after the enlargement of excavation of riverbed dam foundation.

Figure 8. Overall view of foundation surface of Xiluodu arch dam (1:2000).

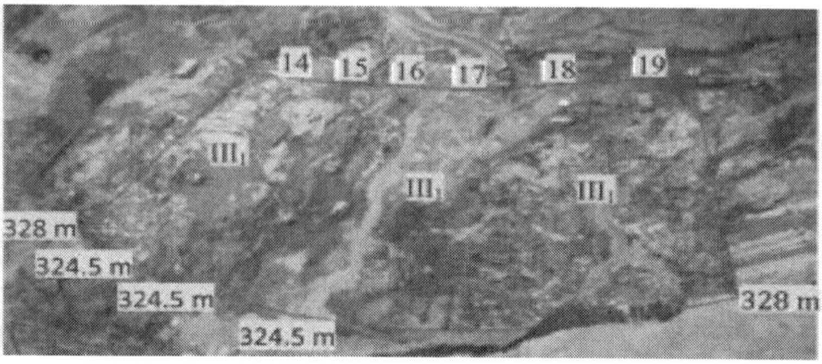

Figure 9. Overall view of riverbed foundation surface after global enlargement of excavation of riverbed dam foundation (1:1000). The numbers 14–19 are the sections.

Global enlargement of excavation (HCEC, 2005, Wang and Yang, 2009, Fan et al., 2012 and Lu et al., 2013) means that all rocks IV_1 and III_2 are completely removed at the foundation elevation over 324.5 m, the inter-laminar and intra-formation dislocation band and the intensively weathered inter-layers are partially replaced through groove carving. All of the replacement excavation is included in the excavation of the dam foundation in order to reduce the effect of secondary excavation of replacement on the construction progress. At the same time, the downstream replacement excavation areas of geological defects should be enlarged adequately to the banks and lower reaches, which are basically 5–10 m in length. This can reduce the adverse effect of sudden change in the geometrical shape of replaced foundation surface on the dam structure.

After the global enlargement of excavation, dam foundation rocks III_1 shared a larger percentage of 86%, and rocks III_2 shared 14.1%. Rocks III_2 inside the dam foundation were distributed in an intercalated and lenticular shape, in poor continuity spatially. Rocks III_2 distributed on the foundation surface were removed through groove carving and foundation cleaning. The quality of rock mass can meet the foundation requirements in a global sense (Wang and Yang, 2009). Fig. 10 displays the overall view of foundation rock mass quality before and after groove carving.

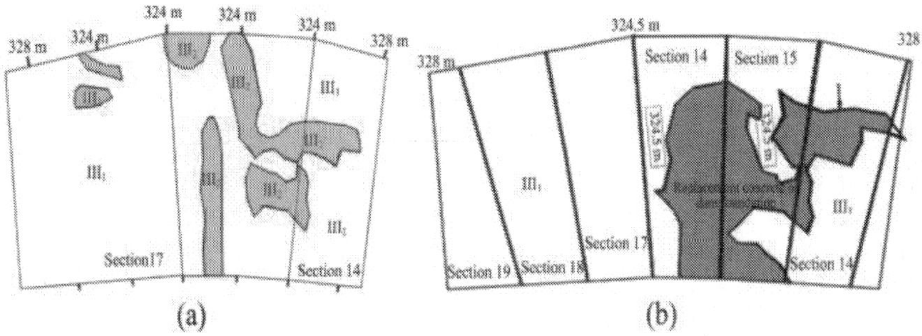

Figure 10. Dam foundation rock mass quality before and after groove carving of riverbed section at elevation of 328–324.5 m. (a) Before groove carving. (b) After groove carving.

Optimization design of riverbed dam bottom section

Sonic wave test data after excavation show that the rocks under the foundation surface were mostly weakly weathered to slightly weathered rocks III_1. The rock mass at shallow depth of 1–6 m was inninmorite, with average wave velocity V_P >4600 m/s, characterized by high rock strength and poor uniformity. Rocks within the depth of 20 m under the foundation surface were rock mass III_2 (about 15%), which were mainly affected by the intersection of intra-formational belts with gentle inclination. The wave velocity is V_P = 2500–4000 m/s with gentle spatial distribution and extension length basically of 10–30 m. The dislocation interfaces were dominated by angular gravels, with little rock debris and no mud. Given that the inter-laminar and intra-formation dislocation bands distributed in the rock mass of the riverbed dam section were roughly perpendicular to the load of arch dam beam, the key parts for foundation deformation treatment, it is desirable to improve the rocks and foundation rigidity by grouting methods. At the same time, the design of bottom structure should be optimized, and the stressed area of the foundation was enlarged so that the foundation deformation and stability requirements could be met (Wang and Yang, 2009). For this purpose, optimization design was conducted for the arch dam foundation structure on the basis of "global enlargement of excavation" concept, with comprehensive treatment measures employed such as global continuous pouring, consolidating grouting, revetment of slope and toe protection, expansion of overall foundation structure, etc. (Wang and Yang, 2009, Wang et al., 2010, Fan et al., 2012 and Lu et al., 2013).

Global continuous pouring was first to use the cast-in-place concrete of the same grade as used in dam at the replacement area below the elevation of 374 m on the left bank and 376 m on the right bank until the riverbed elevation of 324.5 m (except for the sections C3 and C4 using groove carving method) was reached, in combination with the dam concrete. The partings between replacement areas were the transverse joints of arch dam expanding naturally, which could reduce the negative effect of concrete replacement on the construction progress. Then, the back-filled concrete in the extension of excavation region in the downstream dam toe area was used as the dam toe protection, so that it can be cast associated with the dam concrete. Fig. 11 shows the structure joints of replacement areas.

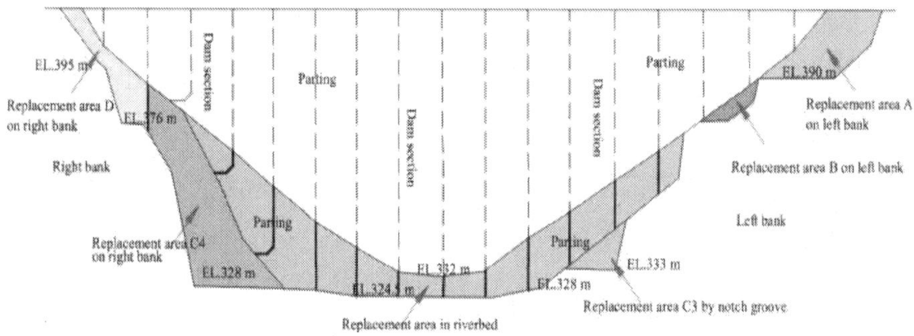

Figure. 11. Joints of foundation replacement area in riverbed dam section.

"Consolidating grouting" was proposed to use grouting to consolidate the rocks below the elevation of 324.5 m, especially the rocks III_2 (15%) within 20 m under the foundation surface. This can improve the foundation integrity and thus control the dam foundation deformation. The first step was to increase the depth of grouting hole. The second was using intensive grouting; the third was to adjust the grouting pressure and the new grouting technology. Thus the grouting water-cement ratio was optimized and grinding fine cement was used. The fourth was to apply anchor piles ($3\phi32$ mm) inside the grouting holes to further improve the integrity of foundation and the dam.

"Revetment of slope and toe protection" was to apply revetment to upstream slope surfaces and toe protection to downstream excavation slope surfaces, respectively. The application of downstream slope toe protection could improve the high stressed distribution state in the dam toe area. The application of upstream slope revetment could prevent any possible cracks propagating toward the dam toe when the upstream contact surface of arch dam was in a tensile state after impoundment.

"Expansion of overall foundation structure" was used for the comparison purpose of the two schemes, i.e. the schemes of "expansion of overall foundation structure" and "cushion for separate foundation abutment", in terms of structural safety and quality control in construction period. The analysis results obtained from the "expansion of overall foundation structure" scheme show that the dam displacement and stress in the construction and operation periods can meet the design requirements. The overload factor of safety was greater than those in the feasibility study and optimization schemes. Thus, this scheme was finally adopted. The crack initiation load of upstream dam toe was $P_1 = (2\text{–}2.5)\ P_0$ (P_0 is the normal load combination, deadweight of dam + upstream normal water level + corresponding downstream water level + silt pressure + temperature decrease) with $K_1 = 2\text{–}2.5$, the nonlinear deformation load was $(5\text{–}6)P_0$ with $K_2 = 5\text{–}6$, and the ultimate load was $P_3 = (9\text{–}9.5)P_0$ with $K_3 = 9\text{–}9.5$.

Foundation reinforcement measures for Xiluodu arch dam

The complicated joint loading features of super-high arch dam determine that the arch dam foundation must have sufficiently global integrity, stability and impermeability, and the rocks in the dam spandrel must be stable (ASCE, 1974 and Lin et al., 2011). To meet these requirements, the foundation reinforcement measures such as consolidating grouting, curtain grouting, foundation drainage, and anchor cable of dam toe were used for Xiluodu arch dam at the construction stage, in combination with the geological conditions revealed after the excavation of riverbed dam section.

(1) Consolidating grouting

To improve the foundation rock integrity and uniformity, consolidating grouting using inclined holes was conducted across the foundation of arch dam, extending 5 m and 10 m respectively toward the upstream area of the dam heel and the downstream area of the dam toe. The methods of non-weight grouting + weighted-tube grouting, grouting with weighted concrete cover, and grouting without concrete cover were applied following the principle of three sequences and step consolidation. After the consolidating grouting, the average wave velocity of rock mass of dam foundation increased, with a significant increase in rocks III_2 and the weak belts. The areas with small wave velocity values were decreased significantly. The comprehensive elastic modulus was increased with an average rate of about 34.5%, the maximum rate of about 62% (the maximum value of 14.5 GPa), and the minimum rate of about 8% (the minimum value of 9.72 GPa). Quality inspection after grouting shows that the rock permeability can meet the design requirements. Quality inspection and physical survey results show that the inter-laminar and intra-formation dislocation bands on riverbed section were significantly enhanced, suggesting the improved integrality and uniformity of rock foundations.

(2) Foundation waterproofing and drainage system

In order to reduce the osmotic pressure of the dam foundation and dam abutments on two banks, comprehensive waterproofing and drainage concept of "prevention first, drainage complementally, and combined prevention and drainage" (Zhou et al., 2006) was employed in association with the water permeability of the dam site and the distribution of main permeable structures. The anti-seepage curtain walls for the dam were arranged in the dam foundation corridor and the 6-layer grouting adits in left and right bank mountains. The upper and lower layer adit curtain walls were connected by lap joint curtain. The dam drainage system consisted of one transverse drainage curtain under the dam foundation, one drainage curtain at the dam toe, and two longitudinal drainage curtains and four transverse drainage curtains in dam abutment on each bank.Fig. 12 shows the arrangement of anti-seepage curtain walls and drainage systems in the dam foundation corridor and on both banks.

(3) Contact grouting of dam foundation

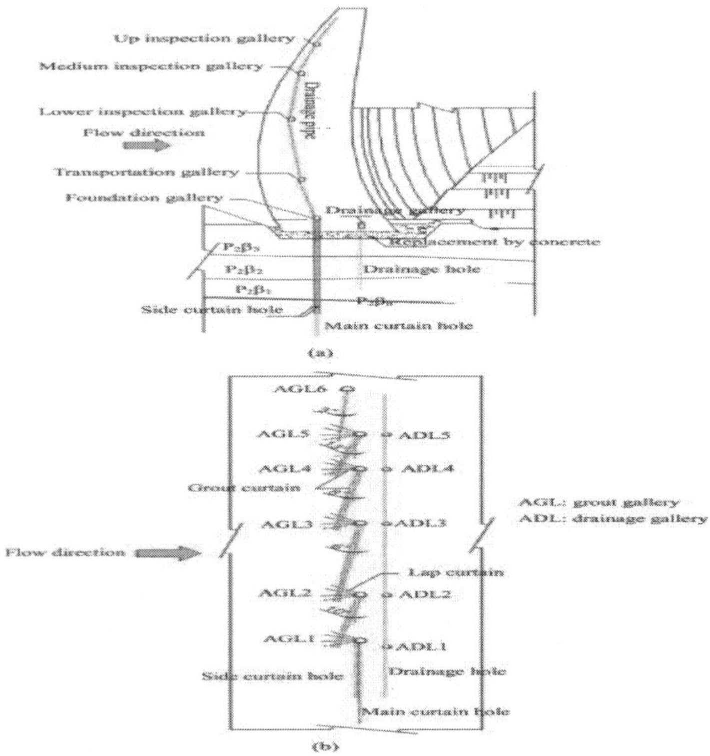

Figure. 12. Arch dam foundation curtain grouting and drainage arrangement system. (a) Anti-seepage curtain of dam foundation. (b) Anti-seepage curtain wall.

Foundation contact grouting was conducted for the steep slopes of dam foundation, the replacement blocks, and the steep slope faces with one-side slope excavation with slope angle greater than 50° and the roofs. The purpose of this grouting was used to increase the shear strength and impermeability of contact surface. This can be done with the method of combined consolidating grouting and curtain grouting.

(4) Anchor cable of dam toe

Anchor cables were arranged for the dam toe for the targeted elevations and sections. Anchor cables were basically arranged along the downstream protection of dam toe, in line with the revetment structure, the direction of which was perpendicular to the foundation surface, inclining upward slightly.

Stability analysis of Xiluodu arch dam foundation

To better understand the mechanical behaviors of arch dam under normal load combinations, the overall stability analysis and failure modes were conducted with 3D geomechanics model test (Zhou et al., 2010) and 3D finite element simulation analysis (Lin et al., 2010).

3D geomechanics model test

Model design

The model with geometric similarity ratio of 250 was used. The dimensions of this model were 155 m (upstream) × 825 m (downstream) × 230 m (height), which are about 0.55, 2.9 and 0.81 times the prototype dam height, respectively. The model height of both sides is 510 m, which is about 1.8 times the prototype dam height. Fig. 13 displays the model design.

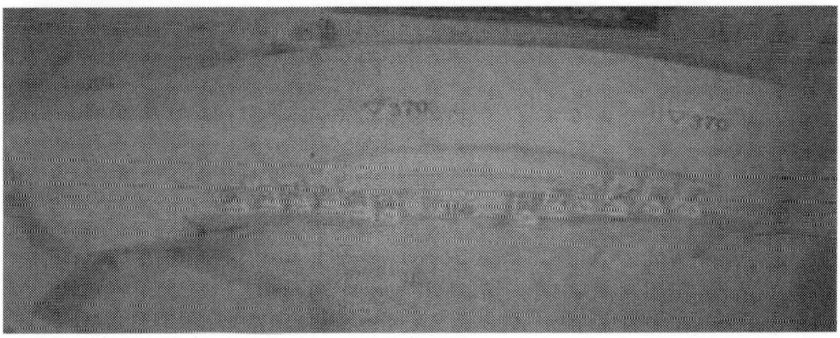

Figure 13. 3D geomechanical model – foundation replacement and dam toe protection (1: 250).

The model was simulated associated with various rock types, weak structural surfaces, 6 layers of inter-laminar and intra-formational dislocation bands (C_2, C_3, C_7, C_8, C_9, $L_{c5}$and L_{c6}) and 2 groups of steep fractures (left-bank fracture N20°–30°W/SW(NE)∠70°–85°, right-bank fracture N60°–80°E/SE (NW)∠65°–85°). The foundation treatment effect and dam toe protections were also simulated. Table 10 presents the physico-mechanical parameters of various rocks. Fig. 14 illustrates the layout of the test measurement system.

Table 10. Physico-mechanical parameters of various rocks.

Rocks		Deformation modulus (GPa)	Volume weight (10^4 kN/m^3)	μ	Shear strength		Shear resistance		Remarks
					c' (MPa)	f'	c (MPa)	f	
Dam concrete		24	2.4	0.167	2.5		0		
II		21.5	2.85	0.2	2.5	1.35		0.99	
III_1	III_{1-1}	14.9	2.85	0.25	2.2	1.25		0.92	Elevation over 560 m
	III_{1-2}	13							Elevation 560–350 m
	III_{1-3}	12							Elevation 350–330 m
	III_{1-4}	9							Elevation 330–300 m
	III_{1-5}	10							Elevation below 300 m
III_2	III_{2-1}	6	2.75	0.28	1.4	1.2		0.84	Elevation over 560 m
	III_{2-2}	5.5							Elevation 560–330 m
	III_{2-3}	5							Elevation below 300 m
IV_1		3	2.6	0.3	1	1.02		0.7	

Note: c' and c are the cohesions, f' and f are the shear strength factors.

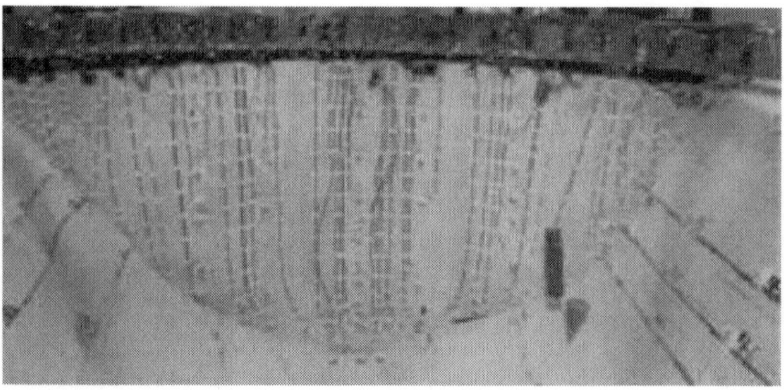

Figure 14. Arrangement of strainmeters on dam face in downstream area of model (1: 250).

Analysis of work behaviors of arch dam under normal load combination

The maximum displacement of the dam along river direction under normal load combination was 125.6 mm, which was observed on the top of the arch crown. The displacement of right arch beam on the top was greater than that of the left arch beam; whilst in the other positions, the displacement of left arch beam was greater than that of the right arch beam. As the rocks beneath the foundation were composed of multiple rock layers that were mixed and the faulted bedding plane was observed associated with developed inter-laminar and intra-formational dislocation band, the overall rigidity of foundation was low, thus the displacement of the lower part of arch abutment was significantly large. The displacement of crown cantilever along river direction basically deformed towards left, and arch abutments deformed towards the mountain. Fig. 15presents the displacement of downstream face of arch dam under the normal load combination.

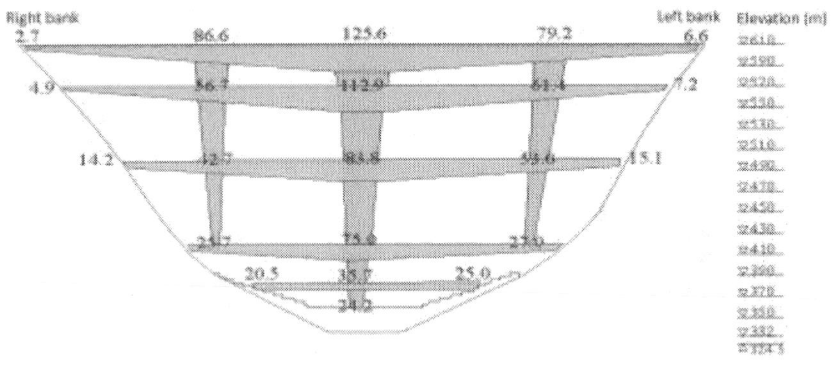

Figure 15. Deformation of downstream face of arch dam under normal load combination (unit: mm).

Under normal load combination, the maximum tensile stress of the upstream face dam heel was 1.57 MPa, and the maximum principal compressive stress of the downstream face was 7.85 MPa, as shown in Fig. 16. Due to the application of toe protection, the compressive stress of the downstream dam toe was lower than those in the feasibility and optimization schemes. The downstream face was in a form of natural arch along the trajectory of the main compressive stress, and the stress was smoothly distributed along the arch. Tensile stress appeared at several positions of the downstream dam face; especially a tensile stress area near the left bank weak face was observed, where tensile cracks appeared as a form of final failure.

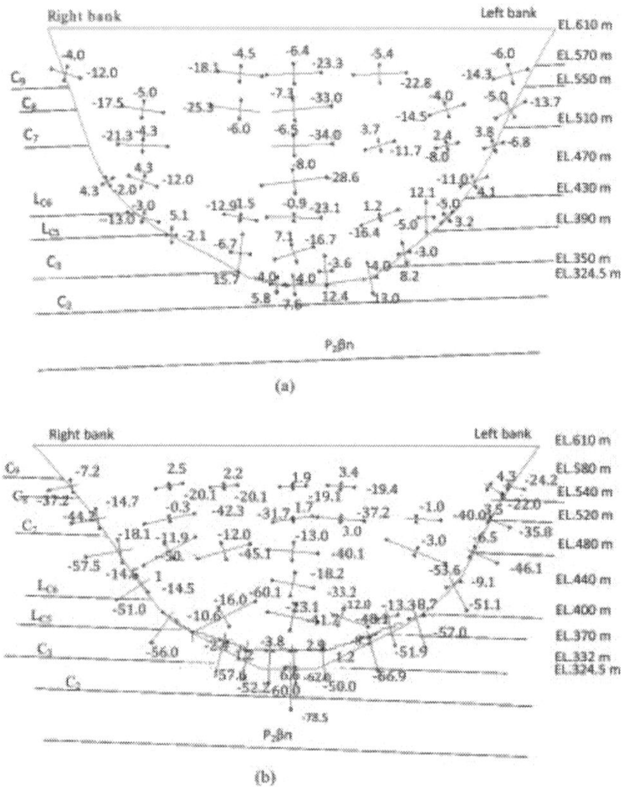

Figure 16. Stress distribution of upstream and downstream faces of arch dam under normal load combination (×0.1 MPa). (a) Upstream face. (b) Downstream face.

Analysis of overall stability of arch dam

The failure mode of arch dam was shown as a result of tensile cracks observed on the upstream dam heel, compression shear on the downstream dam face, and cracking of rocks. No large-scale overall sliding failure was observed on the dam and foundation. When the load of the upstream face was $2P_0$, cracks initiated at the right dam heel, and gradually extended to the high-elevation position till L_{c6} on the right bank. When the load reached $3.5P_0$, the left dam heel was cracked, and at the same time the dam heel on the upstream riverbed was also cracked with a shallow depth. As the load increased, the cracks extended towards both sides. When the load reached $5P_0$, damaged positions increased significantly.

Due to the application of toe protection on the downstream face, cracks began to appear between L_{c6} and L_{c5} on the left bank when the load reached 2.5P_0, and then gradually extended towards the upper and lower sides separately. When the load increased to 3.5P_0, cracks began to appear near Lc6 on the right bank, and extended upwards and downwards the dam. When the load increased to 6P_0, failure was accelerated.

The tensile fracture load of upstream dam heel was P_{1t} = (2–2.5)P_0, and the compressive fracture load of downstream dam toe was P_{1c} = (2.5–3)P_0. The load when significantly large deformation occurred was P_2 = (5–6)P_0, and the ultimate load capacity wasP_3 = (9–9.5)P_0. The application of toe protection below the elevation of 359 m of dam foundation can significantly improve the stress state of the dam, facilitating the upstream dam heel's crack resistance and the downstream dam toe's fracture resistance. Fig. 17shows the failure process of upstream and downstream faces of Xiluodu arch dam.

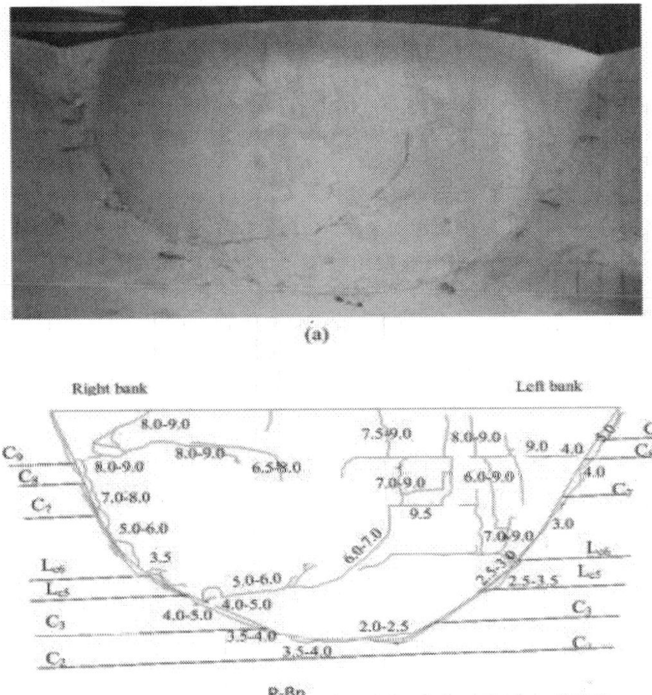

Figure 17. Failure process sketches of upstream and downstream faces of Xiluodu arch dam. (a) Upstream face. (b) Downstream face. The numbers are the overload coefficients.

Rock foundation deformation and failure mechanism

When the load increased to $2P_0$, the upper and lower layers on the upstream side of left and right bank (L_{c6}) began to move. Dam foundation cracking appeared first on the upstream side of the top of right dam abutment, and P_1 was $(3.5–4)P_0$. When the load increased to $(5–6)P_0$, both the dam body and foundation suffered a relatively large deformation, and slippage and cracking took place in the left bank joints and inter-laminar and intra-formational dislocation bands. When the load was $(4–5)P_0$, compression and shear failure occurred in the left bank dam toe of the riverbed dam foundation. When it increased to $(5–6)P_0$, compression and shear failure took place in the right bank dam toe. When P was greater than $8P_0$, left bank cracks opened and propagated downstream rapidly (at elevation of 530–390 m). Fig. 18 illustrates the deformation and failure of arch dam and foundation.

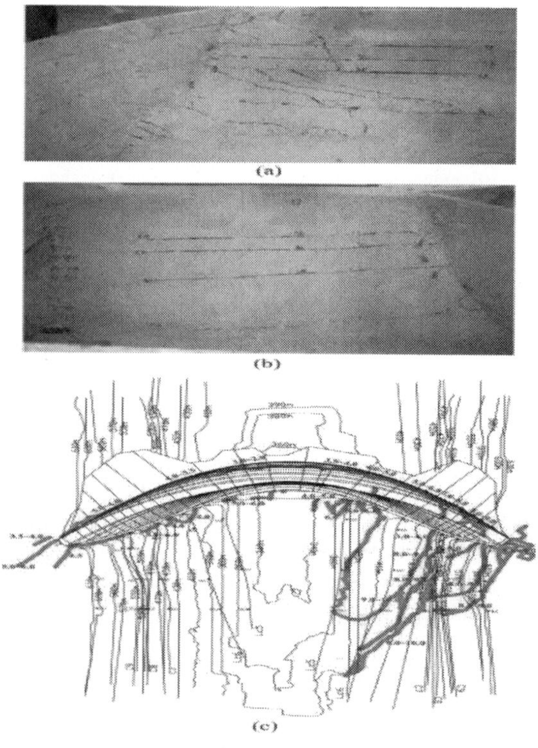

Figure 18. Sketches of deformation and failure of arch dam and foundation during overload. (a) Final failure of rock mass on left bank. (b) Final failure of rock mass on right bank. (c) Sketch of deformation and failure of arch dam and foundation during overload ($P_1 = (3.5–4)P_0$). The numbers colored in blue are overload coefficients.

3D finite element analyses of arch dam

Calculation method and working conditions

The strength of materials was calculated with Drucker–Prager (D–P) yield criterion as

$$f = \alpha I_1 + J_2^{1/2} - H = 0 \quad (1)$$

$$\alpha = \frac{3\tan\varphi}{\sqrt{9+12\tan^2\varphi}} \quad (2)$$

$$H = \frac{3c}{\sqrt{9+12\tan^2\varphi}} \quad (3)$$

where I_1 is the first invariant of stress, J_2 is the second invariant of stress, c is the cohesion, and φ is the friction angle.

The analysis is focused on the working conditions of fundamental scenario that could control the overall safety of the arch dam, i.e. the deadweight of dam body and rock mass, temperature effect, osmotic pressure, the water pressure and silt pressure under normal water level. The working conditions of overload effects are with the same load combination as the fundamental scenario, i.e. the overload is applied through continuous increase of the reservoir water level.

Grid model and materials parameters

Dimensions of the simulations were 1800 m × 1203 m × 597 m, which were determined through the dam axis, i.e. the distances of 1 time dam height upstream, 2 times the dam height on each of the left and right banks, and 2 times the dam height downstream, respectively. The depth was 1 time the dam height beneath the dam foundation. The grids totaled 25,088, and nodes of 28,647. The calculations considered the combination of various rocks, blocks C_9, C_8, C_7, L_{c6}, L_{c5}, C_3, C_2, weak inter-layer $P_2\beta_n$ (strip thickness of 0.3–1.5 m), foundation concrete replacement, and damtoe protection under targeted load combinations. Fig. 19 and Fig. 20 show the simulated replacements of the toe and dam, and inter-laminar and intra-formational dislocation bands, respectively. Table 11lists the parameters of various materials used.

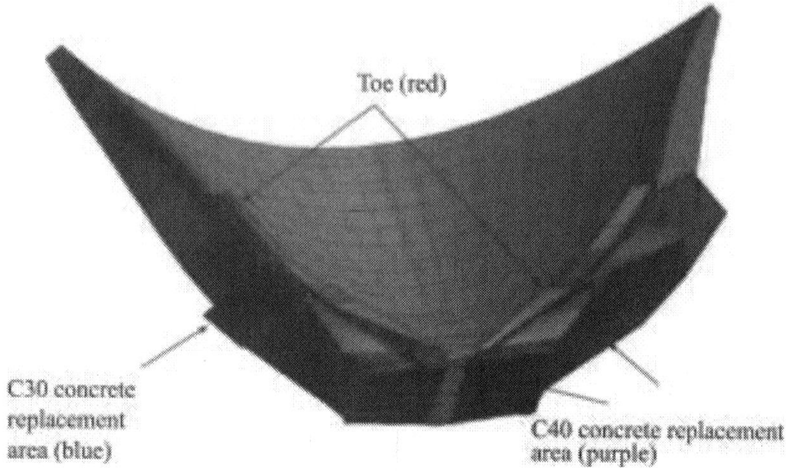

Figure 19. Sketch of relation among foundation concrete replacement, toe protection, and dam.

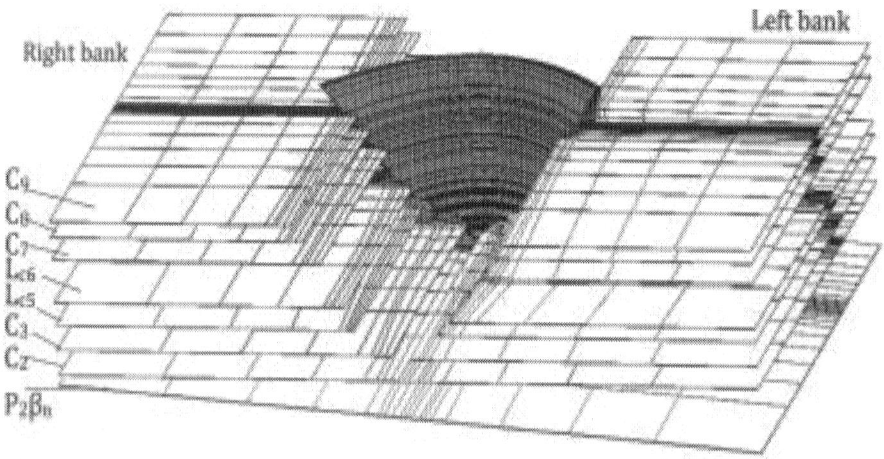

Figure 20. Sketch of the positions of inter- and intra-formational belts.

Table 11. Physico-mechanical parameters of various materials in the dam area.

Material	Density ($g\ cm^{-3}$)	Deformation modulus (GPa)	Poisson's ratio	Shear strength		Shear resistance		Simulated elevation (m)	Simulated thickness (m)
				c' (MPa)	f	c (MPa)	f		
Dam concrete	2.4	24	0.167	5	1.7	0	1		
II	2.85	16.5	0.2	2.5	1.35	0	0.99		
III_1	2.85	11.5	0.25	2.2	1.22	0	0.92		
III_2	2.75	5.5	0.28	1.4	1.2	0	0.84		
IV_1	2.6	3	0.3	1	1.02	0	0.7		
IV_2	2.6	1	0.3	0.5	0.7	0	0.56		
V	2.2	0.5	0.35	0.05	0.35	0	0.3		
Left-bank C_9	2.4	0.5	0.3	0.07	0.4	-	-	563	0.5
Right-bank C_9	2.4	0.5	0.3	0.06	0.4	-	-	562	0.5
Left-bank C_8	2.4	0.8	0.3	0.1	0.44	-	-	513	0.5
Right-bank C_8	2.4	0.9	0.3	0.1	0.44	-	-	520	0.5
C_8	2.4	3	0.3	0.25	0.55	-	-	536	0.5
Left-bank C_7	2.4	1.7	0.3	0.2	0.55	-	-	479	0.6
Right-bank C_7	2.4	1.3	0.3	0.2	0.5	-	-	497	0.6
Left-bank L_{C6}	2.4	0.8	0.3	0.08	0.44	-	-	403	0.6
Right-bank L_{C6}	2.4	0.9	0.3	0.08	0.44	-	-	424	0.6
Left-bank L_{C5}	2.4	0.8	0.3	0.09	0.44	-	-	380	0.6
Right-bank L_{C5}	2.4	0.8	0.3	0.08	0.44	-	-	385	0.6
Left-bank C_3	2.4	0.8	0.3	0.17	0.5	-	-	339	0.5
Right-bank C_3	2.4	0.8	0.3	0.17	0.5	-	-	345	0.5
C_2	2.4	0.5	0.3	0.05	0.35	-	-	302	0.5
$P_2\beta_n$	2.4	1.5	0.3	0.05	0.35	-	-	240	1.8

Analysis of arch dam stress and displacement

Under the normal load combination, the dam displacement was well distributed. The maximum radial displacement of the crown cantilever was 132.5 mm at the elevation of 610 m. The maximum radial displacement of left arch abutment was 27.7 mm, appearing on the downstream face at the elevation of 340 m. The maximum radial displacement of right arch abutment was 25.6 mm on the downstream face at the elevation of 450 m.

The maximum tangential displacement of crown cantilever was 11.5 mm, and the maximum tangential displacement of left arch abutment was 12.2 mm at the elevation of 560 m on the downstream surface. The maximum tangential displacement of right arch abutment was 12.2 mm at the elevation of 520 m on the downstream face. All deformation directions were towards the mountain.

Under normal load combination, the maximum main tensile stress of about 0.17 MPa was observed on the upstream dam face near the left arch abutment at elevation of 610 m. The compressive stress of downstream face increased gradually from the middle part of dam face to the arch abutment, and the high stress area was observed in vicinity of the arch abutment at low and medium elevations. The maximum principal compressive stress was about −16.59 MPa at the elevation of 380 m. The horizontal tensile stress appeared near the elevation of 470 m on the downstream dam face, and the maximum of about 1.13 MPa was found on the right side of the crown cantilever at elevation of 470 m.

Analysis of foundation stress displacement and point factor of safety

(1) Under the working condition of normal load combination, the maximum dam foundation displacement was 25.5 mm observed on the right arch abutment at the elevation 440 m. The difference of the displacements between top elevation and bottom elevation was relatively small, with maximum value not more than 12.5 mm. The displacements of upper and lower surfaces of the dislocation interfaces were relatively small. It can be noted that the rock mass of arch abutment was basically under compression state, and the dam and foundation were in an elastic working state.

(2) Under the working condition of normal load combination, the point factor of safety of arch abutment rock mass was greater than 1.2 on the bank sides, but larger than 1.5 in the deeper part. With consideration of geological faults and weak intercalation, the point factor of safety had the smallest value in the shallow layers on bank slopes at any elevation, but were basically greater than 1.5, meeting the requirement for anti-sliding points of arch abutment (see Table 12). The point factors of safety in the inter-laminar dislocation bands of arch abutment were greater than 1.2 on the bank sides, which were roughly 1.5–2, and greater than 5–15 in deeper part. Their behaviors were all elastic under this working condition (see Table 13). Areas with smaller point factor of safety appeared on the upstream side of the dislocation interfaces, and on the shallow part of bank slope of left-bank C_9, C_7, L_{c6} on the downstream side. Some parts of the arch abutment and right bank L_{c6} and L_{c5} were also reported with low point factor of safety.

Table 12. Point factors of safety of arch abutment.

	Right arch abutment		Left arch abutment	
Elevation (m)	Upstream	Downstream	Upstream	Downstream
610	3	3	3	3
570*	3	3	2	3
540*	1.5	2	1.5	1.5
520*	2	3	1.5	2
500*	2	2	2	2
470	2	2	1.5	1.5
440	2	2	2	2
415*	1.5	1.5	2	2
400*	1.5	1.5	1.5	1.5
380*	2	2	2	2
340*	2	2	2	3
332*	2	2	2	3

Note: "*" means the elevation passing through the fault or weak inter-layers.

Table 13. Point factors of safety of arch abutment in the inter-formational belt.

Belts		Elevation (m)	Point factors of safety
Left bank	C_9	536	Bank side >1.5, deeper part 5–15
	C_8	513	Bank side >1.5, deeper part 5–15
	C_7	479	Bank side >1.5, deeper part 5–15
	L_{C6}	403	Bank side >1.5, deeper part 5–15 (partially yielded)
	L_{C5}	380	Bank side >1.5, deeper part 5–15
	C_3	340	Bank side >2.0, deeper part 5–15
Right bank	C_9	562	Bank side >2.0, deeper part 5–15
	C_8	536	Bank side >1.5, deeper part 5–15 (partially yielded)
	C_7	497	Bank side >2.0, deeper part 5–15
	L_{C6}	424	Bank side >1.5, deeper part 5–15
	L_{C5}	385	Bank side >1.5, deeper part 5–15
	C_3	345	Bank side >2.0, deeper part 5–15

Analysis of dam stability under overload condition

When consolidating grouting for the riverbed dam foundation, revetment to slope and toe protection to the bottom structure, and optimization of the overall structural shapes of bottom structure were considered, the overall overload factors of safety of the arch dam can be calculated. The crack initiation overload coefficient was $K_1 = 2$, the nonlinear deformation overload coefficient was $K_2 = 3.5–4$, and the ultimate overload coefficient was $K_3 = 7–8$. In the conditions of the K_1 and K_2, the stress, deformation, thrust and other indicators of dam foundation were equal to those in the feasibility study scheme, and appeared to be larger than those in the optimization scheme. In the condition of K_3, the overload capacity was equal to that in the optimization scheme, and slightly lower than that in the feasibility study scheme. Under the condition of high water level, the displacement and stress matched with the criteria requirements.

MONITORING DATA ANALYSIS OF FOUNDATION DURING THE CONSTRUCTION AND INITIAL IMPOUNDMENT PERIODS

During the construction and initial impoundment periods of Xiluodu arc dam, the dam deformation recorded showed that the magnitude of the deformation is controllable when water level reached 560 m. The seepage flow increased slowly without abrupt changes with the upstream water level increasing. The osmotic pressure was relatively small in a global sense, which satisfied the design requirements. The compressive pressures of dam foundation and joints of foundation rock were varied regularly and reasonably in a small range. The dam and foundation were in a good working condition.

Analysis of arch dam foundation deformation

Dam foundation deformation

The settlement distribution curves of dam foundation showed that the deformation difference between adjacent dam sections was basically less than 1.8 mm. In the period of concrete cast-in-place of the dam, the foundation suffered a greater deformation with elapsed time under the action of deadweight. During impoundment, only the dam sections of foundation surface outlets suffered slight increase in deadweight when the joint grouting reached the elevation of 563.25 m. With the dam arch effect formed, the upward changes of foundation deformation associated with initial impoundment were minor. When water level reached 560 m, the maximum deformation of foundation was 31.21 mm observed in dam section 16, an increase of 4.67 mm when the initial water level reached 540 m in June 2013. Fig. 21 presents the typical hydrograph of the relevant settlements at typical measuring points during impoundment.

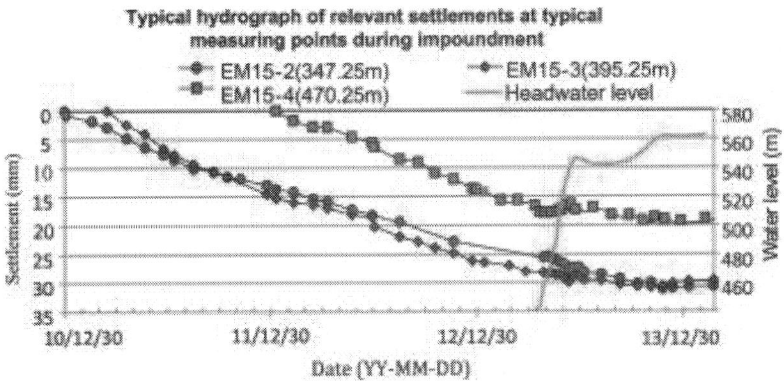

Figure 21. Typical hydrograph of relevant settlements at typical measuring points during impoundment.

Joint aperture of dam foundation

There were 50 m embedded in the dam sections 3–29 and in the concrete replacement areas of Xiluodu arch dam foundation, for the purpose of measuring rock aperture between the base rock surface and the dam bottom surface. At present, the accumulative deformation recorded from the base rocks is −3.82 to 2.38 mm, which was considered to be compressive. In terms of deformation distribution, the compressive deformation of riverbed dam section was larger than that of the bank slope dam section, which was illustrated in a form of greater surplus thrust. The largest compressive deformation was recorded on the upstream side of dam section 17.

After impoundment, the imposed forces of upstream water increased. The dam foundation joints on the upstream side were open deforming, but those in the middle part and on the downstream side were under compression. Moreover, the compressive stress on the downstream side was greater than that in the middle part, which was most likely attributed to the increase in upstream water level. The riverbed dam sections (14–19) tended to be under tension at the upstream dam toe, with the maximum aperture of 0.31 mm recorded in dam section 14. The middle and downstrcam parts of the dam foundation were under compression, with the maximum downstream-side compressive deformation of 0.69 mm recorded in dam section 10 (see Fig. 22).

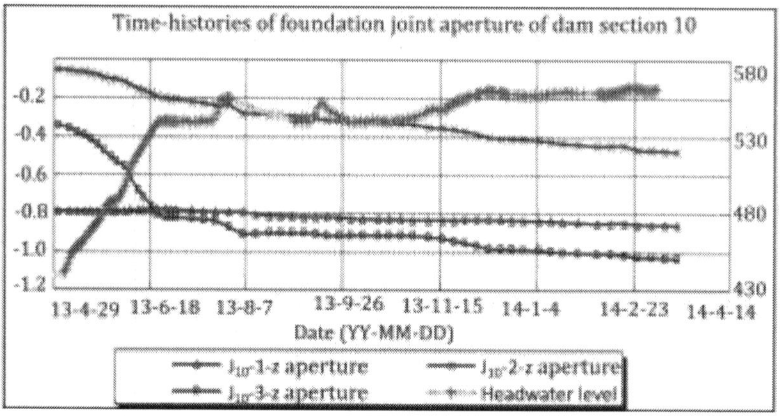

Figure 22. Time-histories of foundation joint aperture of dam section 10.

Osmotic pressure monitoring

During impoundment, the osmotic pressure behind the dam curtain decreased significantly with the elevated water level. The measured water head of osmotic pressure behind the curtain was 0–51.81 m (dam section 15) with increment of water head of −1.41 to 24.63 m. The osmotic pressure coefficient a_1 was 0.06–0.23, and a_2 was 0–0.08, meeting the requirements of design and criteria (a_1 is the uplift coefficient, and a_2 is the residual uplift coefficient).

Analysis of dam foundation stress state

The compressive stress of arch dam foundation was basically 0.19–7.34 MPa before impoundment, and varied from −0.06 to 2.45 MPa after impoundment. Due to the increasing forces applied by upstream water, the compressive stress of dam foundation on the upstream side tended to decrease after impoundment, by a maximum value of 1.14 MPa recorded at the elevation of 343.65 m of the dam section 12. The compressive stress of dam foundation on the downstream side tended to increase by a maximum value of 2.45 MPa recorded in dam section 10. In the same dam section, the decrease of upstream-side compressive stress was visibly equal to the increase of downstream-side compressive stress. The changing regulation of compressive stress of dam foundation was consistent with the changing rule of joint aperture of dam foundation. Fig. 23 and Fig. 24 present the distribution of compressive stress of dam foundation and time-histories of compressive stress of typical dam sections, respectively.

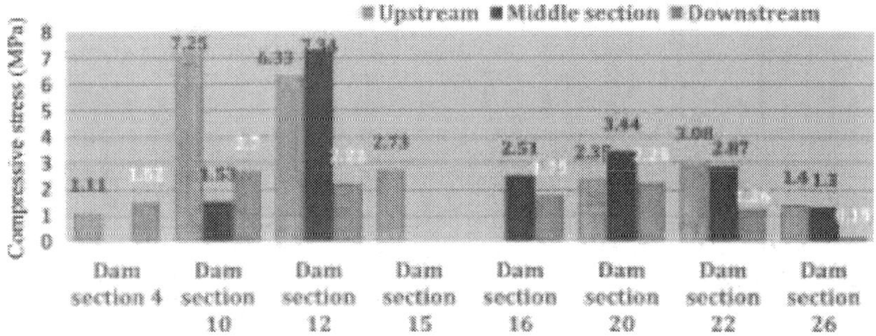

Figure 23. Distribution of compressive stress of arch dam foundation.

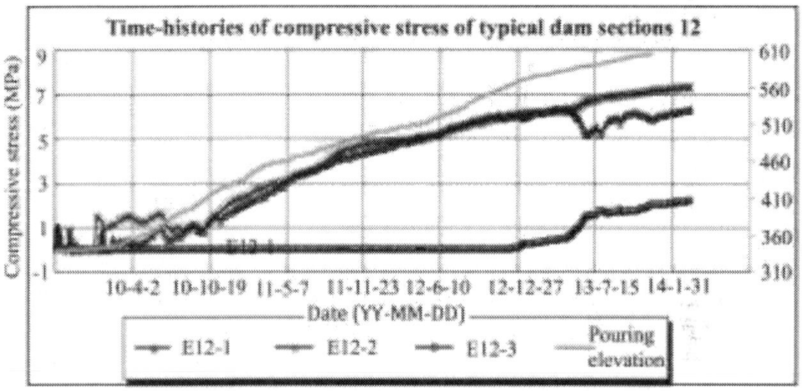

Figure 24. Time-histories of compressive stress of typical dam sections 12.

Back analysis of monitoring results

For the purpose of understanding the overall performances of the dam, timely detection of various possible cracking risks during impoundment is necessary in terms of back analysis according to the monitoring data of the dam, which can facilitate impoundment arrangement in the next step. Therefore, the ACM was used to conduct feedback analysis using the monitoring data when the impoundment reached 540 m, in association with the subsequent impoundment plan. The key predication indicators for the arch crown beams in the condition of impoundment up to 560 m are: (1) radical displacement will be (7.76 ± 3) mm to (21.56 ± 5) mm; (2) vertical displacement will be (-3.03 ± 2) mm to (29.06 ± 3) mm (Liu et al., 2014). When the water level reached 560 m, the monitoring data of the crown cantilever at different elevations were all within the design values.Table 14 gives the vertical and radial displacements of dam section 15.

Table 14. Vertical and radial displacements of dam section 15.

Location	Vertical displacement (mm)		Radial displacement (mm)	
	Predicted value	Measured value	Predicted value	Measured value
Corridor at elevation 347 m	28.38 ± 3	30.95	7.76 ± 3	6.71
Corridor at elevation 395 m	29.06 ± 3	30.93	17.06 ± 3	13.81
Corridor at elevation 470 m	15.93 ± 3	20.23	21.56 ± 5	15.57
Corridor at elevation 527 m	1.13 ± 2	2.82	16.85 ± 3	8.02
Corridor at elevation 563 m	−3.03 ± 2	2.51	14.39 ± 3	3.49

CONCLUSIONS

The optimization design of Xiluodu super-high arch dam is a complicated task, which is closely related to the site-specific geological conditions. Thus a process of dynamic optimization adjustment and timely safety evaluation is desirable. This optimization process requires the introduction of careful survey evaluation, key technologies, excavation optimization, body shape modification and adequate structural design of arch dam, dam performance evaluation, adequate simulation analysis tools and/or geomechanical model test, feedback analysis of monitoring data, etc.

1) At the feasibility study stage, the foundation surface of Xiluodu arch dam was considered to be mainly placed on the slightly weathered to fresh rocks II.
2) At the optimization stage, the weakly weathered rocks III_1 and III_2 were used as foundation surface of the arch dam, which is reasonable and feasible through common analysis method, 3D geomechanics model test, and 3D finite element numerical analysis. Compared with the feasibility study scheme, the dam concrete volume was decreased by 1.1 million m^3 approximately, and the foundation excavation volume was decreased by around 1.61 million m^3. The maximum excavation slope height on left and right banks was

decreased by 40 m, with investments of nearly RMB 600 million saved. In addition, construction period can be shortened by 3 months.

3) At the technical construction stage, according to the site-specific geological conditions revealed in riverbed dam foundation after excavation, the foundation excavation shape and foundation treatment requirements were adjusted in time, and the bottom foundation structure was optimized through the concepts of "overall extension of excavation + consolidating grouting + overall structure integrity + continuous cast-in-place concrete". The foundation and dam body works well. Through 3D geomechanics model test and 3D finite element analysis, the arch dam is considered to have a good dam stress, excellent overall stability and large overload capacity. This is also verified by the monitoring data in the construction period and the initial impounding and operation periods.
4) In view of the construction practices of Xiluodu arch dam, it is feasible to use rocks III_1 and III_2 as the foundation surface of super-high arch dam. But desirable reinforcement measures for the dam foundation must be taken. The structural optimization design is considered for the arch dam foundation by applying the forces on the arch dam to form a global dam foundation. At the same time, attentions should be paid to the geological conditions of dam section on riverbed. Thus advanced geological survey should be conducted prior to construction.

CONFLICT OF INTERESTS

The authors wish to confirm that there are no known conflicts of interest associated with this publication and there has been no significant financial support for this work that could have influenced its outcome.

ACKNOWLEDGMENTS

The authors would like to thank the Chengdu Engineering Corporation Limited of Power Construction Corporation of China, Tsinghua University, and other Companies to provide the relevant results contributed to this paper

REFERENCES

1. American Society of Civil Engineers (ASCE). Foundations for dams. New York: American Society of Civil Engineers; 1974. p. 7e21.
2. Chengdu Hydroelectric Investigation & Design Institute of State Power Corporation (CHIDI). Final report on engineering design technology of Ertan hydropower station on Yaruo River. Chengdu, China: Chengdu Hydroelectric Investigation & Design Institute of State Power Corporation; 2000 (in Chinese).
3. Department of the Interior Bureau of Reclamation, USA (DIBR). Design criteria for concrete arch and gravity dams. United States: Department of the Interior Bureau of Reclamation; 1977.
4. DL/T5346-2006. Design specification for concrete arch dams. Beijing: China Electric Power Press; 2007 (in Chinese).
5. Fan Qixiang, Zhou Shaowu, Li Bingfeng. Key technologies of rock engineering for construction of Xiluodu super-high arch dam. Chinese Journal of Rock Mechanics and Engineering 2012;31(10):1998e2015 (in Chinese).
6. HydroChina Chengdu Engineering Corporation (HCEC). Optimized design report of concrete arch dam of Xiluodu hydroelectric project on Jinsha River. Chengdu: HydroChina Chengdu Engineering Corporation; 2005 (in Chinese).
7. Li Zan, Chen Fei, Zheng Jianbo. Super-high arch dam hub analysis and key issue study. Beijing: China Electric Power Press; 2004. p. 1 (in Chinese).
8. Lin Peng, Zhou Weiyuan, Yang Qiang, Yang Ruoqiong. The overall stability analysis and safety evaluation report of dam-foundation of Xiluodu arch dam. Beijing: Tsinghua University; 2010 (in Chinese).
9. Lin Peng, Wang Renkun, Kang Shengzu, Zhang Haichao, Zhou Weiyuan. Key problems of foundation failure, reinforcement and stability of superhigh arch dams. Chinese Journal of Rock Mechanics and Engineering 2011;30(10):1945e58 (in Chinese).
10. Liu Dawen, Chen Xugao, Shao Naichen, Cai Dewen. Feedback analysis report of arch dam monitoring during initial impound of Xiluodu hydroelectric project on Jinsha River. Chengdu: HydroChina Chengdu Engineering Corporation; 2014 (in Chinese).
11. Liu Shihuang. Optimization of foundation surface of Laxiwa arch dam. Beijing: Hydropower Geology and Survey; 1996. p. 17e21 (in Chinese).
12. Lu Youmei, Fan Qinxiang, Zhou Shouwu, Li Bingfeng. Key technologies for construction of Xiluodu high arch dam on Jinsha River. Journal of Hydraulic Engineering 2013;32(1):187e95 (in Chinese).
13. SL282-2003. Design criteria for concrete arch dam. Beijing: China Water Power Press; 2003 (in Chinese).

14. SNiP 2.06.06-85. Design specification of concrete and reinforced concrete dam. Moscow. 1985.
15. Wang Qihong, Wang Miao, Ma Yile. Standard determination of usable rock mass of high arch dam. Journal of North China Institute of Water Conservancy and Hydroelectric Power 2010;31(4):123e6 (in Chinese).
16. Wang Renkun. Analysis and evaluation on optimization of design for foundation interface of super high arch dam. Ph.D. Thesis. Beijing: Tsinghua University; 2007 (in Chinese).
17. Wang Renkun, Yang Jianhong, Cui Changwu, Zhang Shaocheng, Chen Kui. Special research report of dam foundation treatment measures for Xiluodu hydroelectric project on Jinsha River (elevation of 510e610 m). Chengdu: HydroChina Chengdu Engineering Corporation; 2007a (in Chinese).
18. Wang Renkun, Yang Jianhong, Cui Changwu, Zhang Shaocheng, Chen Kui. Special research report of dam foundation treatment measures for Xiluodu hydroelectric project on Jinsha River (elevation of 400e510 m). Chengdu: HydroChina Chengdu Engineering Corporation; 2007b (in Chinese).
19. Wang Renkun, Yao Dewu, Chen Liping, Zhang Shaocheng. Report of rock mass geophysical prospecting testing results for dam foundation of Xiluodu hydroelectric project on Jinsha River. Chengdu: HydroChina Chengdu Engineering Corporation; 2007c (in Chinese).
20. Wang Renkun, Lin Peng. Analysis and evaluation of optimizing design for foundation excavation of Xiluodu super-high arch dam. Chinese Journal of Rock Mechanics and Engineering 2008;27(10):2010e8 (in Chinese).
21. Wang Renkun, Yang Jianhong, Cui Changwu, Zhang Shaocheng, Chen Kui. Special study on dam foundation treatment measures for Xiluodu hydroelectric project on Jinsha River (elevation below 400 m). Chengdu: HydroChina Chengdu Engineering Corporation; 2008 (in Chinese).
22. Wang Renkun, Yang Jianhong. Special design report of arch dam bottom structure of Xiluodu hydroelectric project on Jinsha River. Chengdu: HydroChina Chengdu Engineering Corporation; 2009 (in Chinese).
23. Yang Jianhong, Cui Changwu, Zhang Shaocheng, Chen Kui. Special research report of evaluation of geological conditions of dam foundation engineering for Xiluodu hydroelectric project on Jinsha River (elevation of 510e610 m). Chengdu: HydroChina Chengdu Engineering Corporation; 2007a (in Chinese).
24. Yang Jianhong, Cui Changwu, Zhang Shaocheng, Chen Kui. Special research report of evaluation of geological conditions of dam foundation engineering for Xiluodu hydroelectric project on Jinsha River (elevation of 400e510 m). Chengdu: HydroChina Chengdu Engineering Corporation; 2007b (in Chinese).

25. Zhou Weiyuan, Yang Ruoqiong, Lin Peng, Yang Qiang. 3D geomechanics model test report of Xiluodu arch dam. 1st ed. Beijing: Tsinghua University; 2004 (in Chinese).
26. Zhou Weiyuan, Yang Ruoqiong, Yang Qiang, Lin Peng. Research report of Xiluodu arch dam based on 3D FEM. Beijing: Tsinghua University; 2006 (in Chinese).
27. Zhou Weiyuan, Yang Ruoqiong, Yang Qiang, Lin Peng. 3D geomechanics model test report of Xiluodu arch dam. 2nd ed. Beijing: Tsinghua University; 2010 (in Chinese).
28. Zhu Bofang. Summary of topics of international seminar on arch dam. Hydropower 1988;8:49e52 (in Chinese).

CITATION

Qixiang Fan, Shaowu Zhou, Ning Yang, Optimization design of foundation excavation for Xiluodu super-high arch dam in China, Journal of Rock Mechanics and Geotechnical Engineering, Volume 7, Issue 2, April 2015, Pages 120-135, ISSN 1674-7755, http://dx.doi.org/10.1016/j.jrmge.2015.03.001.

CHAPTER 7

Numerical Analyses in the Design of Umbrella Arch Systems

J. Oke, N. Vlachopoulos, M.S. Diederichs

ABSTRACT

Due to advances in numerical modelling, it is possible to capture complex support-ground interaction in two dimensions and three dimensions for mechanical analysis of complex tunnel support systems, although such analysis may still be too complex for routine design calculations. One such system is the forepole element, installed within the umbrella arch temporary support system for tunnels, which warrants such support measures. A review of engineering literature illustrates that a lack of design standards exists regarding the use of forepole elements. Therefore, when designing such support, designers must employ complex numerical models combined with engineering judgement. With reference to past developments by others and new investigations conducted by the authors on the Driskos tunnel in Greece and the Istanbul metro, this paper illustrates how advanced numerical modelling tools can facilitate understanding of the influences of design parameters associated with the use of forepole elements. In addition, this paper highlights the complexity of the ground-support interaction when simulated with two-dimensional (2D) finite element software using a homogenous reinforced region, and three-dimensional (3D) finite difference software using structural elements. This paper further illustrates sequential optimisation of two design parameters (spacing and overlap) using numerical modelling. With regard to capturing system behaviour in the region between forepoles for the purpose of dimensioning spacing, this paper employs three distinctive advanced numerical models: particle codes, continuous finite element models with joint set and Voronoi blocks. Finally, to capture the behaviour/failure ahead of the tunnel face (overlap parameter), 2D axisymmetric models are employed. Finally, conclusions of 2D and 3D numerical assessment on the Driskos tunnel are drawn. The data enriched case study is examined to determine an optimum design, based on the proposed optimisation of design parameters, of forepole elements related to the site-specific considerations.

INTRODUCTION

As a design of underground excavations becomes larger and more complex, numerical analysis is required to combat increasingly difficult ground conditions, under which reinforcement may be required prior to excavation (pre-support). Due to its time and cost effectiveness in comparison with other pre-support methods (ground freezing, jet grouted columns, or pipe jacking), the umbrella arch method is increasing in popularity (Volkmann and Schubert, 2007). A corresponding increase in understanding of the interactions between the support system and the surrounding ground is required (Volkmann, 2003). Since 1991, literature has agreed that currently limited level of understanding is due to the lack of objective design criterion for the umbrella arch (Carrieri et al., 1991, Hoek, 1999, Volkmann, 2003, Kim et al., 2005, Volkmann et al., 2006, Volkmann and Schubert, 2006a, Volkmann and Schubert, 2006b, Volkmann and Schubert, 2007, Volkmann and Schubert, 2010, FHA, 2009, Hun, 2011 and Peila, 2013). To aid design, Oke et al. (2014a) arranged the umbrella arch methods into thirteen sub-categories and associated them to applicable, specific failure mechanisms within the umbrella arch selection chart (UASC). This paper focuses on two of the sub-categories which employ the forepole element (confined and grouted in place) of the umbrella arch. It also illustrates the use of numerical modelling with regard to the overall response of the umbrella arch with forepole elements and the optimisation of selected design parameters for a squeezing failure mechanism, as illustrated in Fig. 1. Explicit numerical modelling for the optimisation of forepole element employed in other failure mechanisms, such as anisotropic conditions, is outside the scope of this paper. Investigation in the overall response of the forepole element was conducted using calibrated numerical models of two different tunnelling projects with in-situ data: the Driskos tunnel project (Vlachopoulos, 2009), and the Istanbul metro (Yasitli, 2013). Optimisation of selected design parameters was carried out for the severe squeezing ground at the Driskos tunnel at section 8 + 746 (Vlachopoulos and Diederichs, 2014).

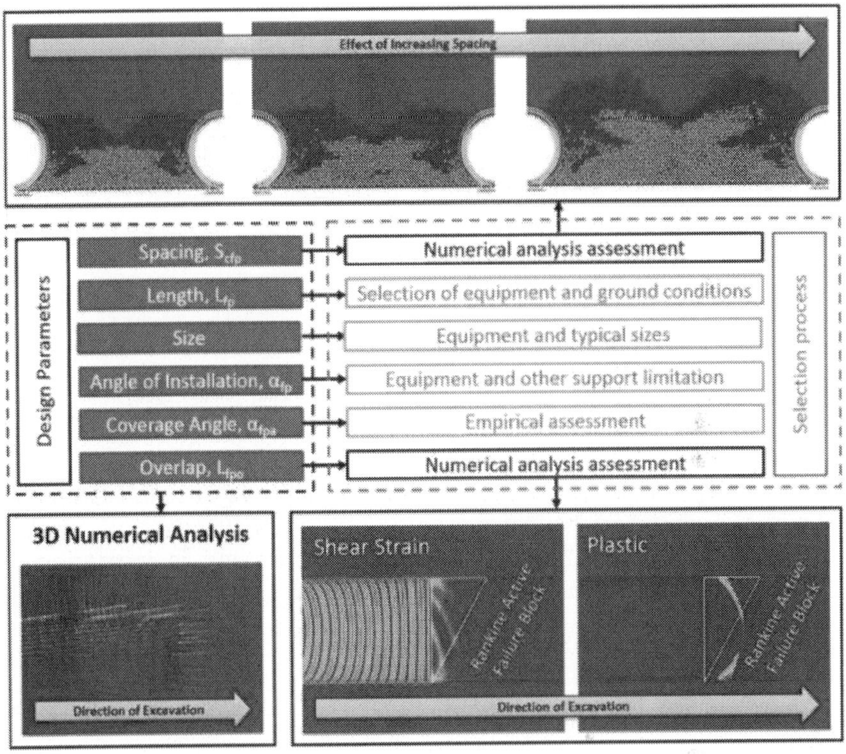

Figure 1. Applications of numerical analysis for design parameters illustrated in Fig. 2, and overall response of forepole umbrella arch methods.

BACKGROUND

Oke et al. (2014a) defined the umbrella arch as a pre-support installed within the tunnel, prior to the first pass of excavation, above and around the crown of the tunnel face which can be made up of spiles (length smaller than the height of excavation), forepoles (length greater than the height of excavation), or grout elements. This paper focuses specifically on the forepole elements, which are part of the temporary support system (e.g. shotcrete, steel sets, rockbolts, as shown in Fig. 2). However, prior to illustrating the extensive three-dimensional (3D) numerical analysis of forepole design, further details are necessary for the design parameters associated with the forepole umbrella arch, relevant investigations of cited literature that highlights important design considerations, the use of two-dimensional (2D) numerical investigation, and their disadvantages.

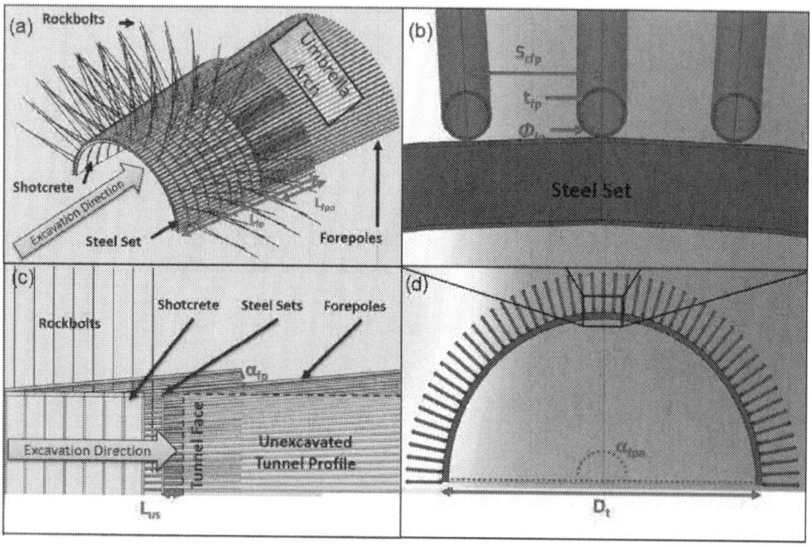

Figure. 2. Structural layout of the umbrella arch temporary support system with forepole elements. Red arrows and text indicate key design dimensions. (a) View of oblique support layout. L_{fp} is the length of forepole, and L_{fpo} is the length of forepole (or umbrella arch) overlap. (b) Viewport of (d). S_{cfp} is the centre to centre spacing of the forepole elements, t_{fp} is the thickness of the forepole element, and ϕ_{fp} is the outside diameter of the forepole element. (c) Profile view of support layout. α_{fp} is the installation angle of the forepole element, and L_{us} is the length of the unsupported span. (d) Cross-sectional view. D_t is the diameter of the tunnel, and α_{fpa} is the coverage angle of the forepole elements.

DESIGN PARAMETERS

The design parameters for the forepole umbrella arch are shown in red in Fig. 2. Fig. 2a displays the length of forepole element (L_{fp}), and the length of forepole (or umbrella arch) overlap (L_{fpo}). The parameter L_{fp} cannot be optimised through numerical analysis as too many non-geomechanical factors governing the design exist. The L_{fp} depends on economic considerations, accuracy of drilling, accessibility of equipment and drillability with respect to ground conditions. The L_{fpo} can be optimised by using relevant numerical modelling. This overlap is required to ensure stability of the system and ground response, as illustrated in Fig. 1. In order to be effective in the longitudinal direction, the embedding of the forepole element requires sufficient distance (length) from the disturbed ground region. This embedment ensures that there will be sufficient longitudinal arching, which is the transfer of stresses at the tunnel face to the support system (in front of the face) and to the stable ground (ahead of the face), as illustrated inFig. 3b. These parameters will be explained further in subsequent sections.

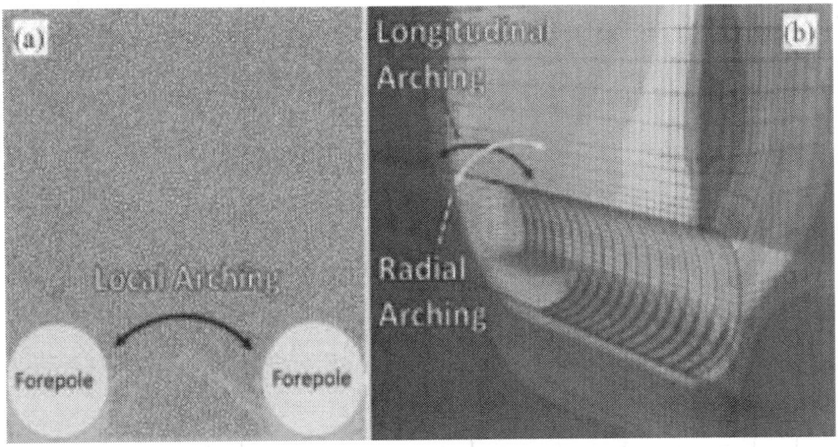

Figure 3. Illustration of arching. (a) Local arching (modified after Doi et al. (2009)). (b) Longitudinal arching and radial arching.

Fig. 2b illustrates the centre to centre spacing of the forepole elements (S_{cfp}), thickness of the forepole element (t_{fp}), and the outside diameter of the forepole element (ϕ_{fp}). The maximum S_{cfp} is defined by the requirement of developing a local arching effect, as shown in Fig. 3a (Volkmann and Schubert, 2007). This local arching can be analysed and captured with numerical models, as illustrated in the top portion of Fig. 1. It is important to note that the FHA (2009) has commented on the occurrence of a common misjudgement of the longitudinal (overestimation) and radial effects (underestimation) of the forepole design. Thus, there is a requirement for analyses on both a local (arching between forepole elements) and a global (complete system response) scale. The size of the forepole element is defined by two parameters: t_{fp} and ϕ_{fp}. Ultimately, these parameters will define the stiffness of the forepole as well as the loading area. This paper will illustrate that numerical modelling can be effective in determining an optimum size of the forepole elements within an umbrella arch arrangement. This optimum size, however, is further influenced by the installation equipment and the commercially standardised elements (pipes) available.

Fig. 2c displays the installation angle (α_{fp}) of the forepole element and the length of the unsupported span (L_{us}). The α_{fp} for spile element within umbrella arch methods can range from 5° to 40° as it is designed to lock in structural components or to ensure a certain thickness of grout barrier around the excavation. For forepole elements, however, the α_{fp} is defined by other temporary support elements (thicknesses of shotcrete and steel sets) as well as equipment clearances, to allow for the minimum possible angle. One must remember that the forepole elements and the umbrella arch support system are not employed in isolation and are used in conjunction with other support elements. The minimum possible angle of installation is deemed to be ideal as most cases result in failure of the ground material up to the forepole elements. When such failure occurs, a niche (or saw-tooth) profile is created, as shown in Fig. 4. This niche profile results in an increasing excavation size which consequently increases the requirement for more and/or larger other temporary support elements, neither of which is economical. The parameter L_{us} is typically defined by the steel sets spacing, although it is also important to take tunnel face stability into consideration when determining the L_{us}.

Figure 4. Niche (saw-tooth) profile construction due to the installation of umbrella arch with forepole elements.

Fig. 2d defines the coverage angle of the forepole elements (α_{fpa}). The tunnel diameter,D_t, is a design parameter that defines the difference between a forepole element and a spile element, as previously mentioned.

The α_{fpa} is defined by the failure mechanism more than the mechanical response of the system. For gravity driven failures, the forepole element only requires a α_{fpa} around the crown (~120°) of the excavation face to protect the workers working underneath. For subsidence driven failure mechanisms, it is more common to employ 180° coverage above the face of the tunnel (as shown inFig. 2d). Similarly, Song et al. (2013) stated that 120° is optimal for weathered rock and 180° for soil. For squeezing ground conditions, Hoek (2001) suggested an increase of the coverage angle from 120° to 180° for severe to very severe squeezing conditions, respectively.

LITERATURE INVESTIGATIONS

While much of the previously discussed literature related to parameters is based primarily on empirical data, the following review sections summarise past work based primarily on numerical modelling. Design parameters based on modelling are summarised in Table 1.

Table 1. Literature review summary of numerical parametric investigations.

Literature numerical analysis investigations	Centre to centre spacing, S_{cfp}	Thickness, t_{fp}	Diameter, ϕ_{fp}	Angle of installation, α_{fpa} (°)	Lateral earth pressure coefficient, K_0	Overburden (m)	Ground material
Volkmann and Schubert (2006a)	30 pieces of 114.3 mm × 6.3 mm forepoles had similar effects of reducing face and maximum settlement when compared to 20 pieces of 139.7 mm × 8.0 mm forepoles			-	-	-	-
Song et al. (2013)	40–60 cm	-	60–114.3 mm	20–180	0.4–0.6	10–40	-
Kim et al. (2005)	-	-	-	-	-	(0.5–3.0)D_t	Weathered soil – weathered rock

Volkmann and Schubert (2006a)

Volkmann and his associated collaborators have carried out laboratory, in-situ, and numerical tests on the forepole element providing either commentary or analysis related to: cost comparison, sizes, length of forepole (L_{fp}), impact of in-situ measurement, disturbed ground foundation, length of forepole (or umbrella arch) overlap (L_{fpo}), and forepole installation methods. Due to the numerical focus of this paper, only the results of quantity, cost and sizes of the forepole elements will be further illustrated.

Volkmann and Schubert (2006a) described the passive response of the support, which requires displacement to mobilise the support effects. The stiffness (t_{fp} and φ_{fp}) of the support element requires less displacement to mobilise the support effects. The analysis of Volkmann and Schubert (2006a), however, did not take into consideration the local arching failure that could occur due to the increase of the spacing of forepole elements. The results of Volkmann and Schubert (2006a) and the arching statements from FHA (2009) reinforce the importance of a design process that incorporates the local and global responses of the system.

Song et al. (2013)

Song et al. (2013) carried out numerical and analytical analyses of a large-diameter steel-pipe-reinforced umbrella arching model with the parameters listed in Table 1. All of the analyses were compared to the factors of safety for bending (FOS_b) and shear (FOS_s). It must be noted that the other supports simulated in this numerical model were not changed (i.e. the shotcrete thickness was constant), and the effect of failing to consider this will be explained in subsequent sections.

The parametric analysis by Song et al. (2013) drew the following conclusions: (i) as ϕ_{fp}increased, FOS_s and FOS_b also increased; (ii) as the overburden depth increased, $FOS_s$and FOS_b decreased and converged; (iii) as S_{cfp} increased from 40 cm to 60 cm, $FOS_s$and FOS_b decreased; (iv) as the Young's modulus of the ground decreased, FOS_s andFOS_b also decreased. On the whole, the FOS_b was found to be more critical than FOS_s. The observations made by Song et al. (2013) suggest that FOS_b should be used as the primary indicator when evaluating the stability of the forepole elements.

Kim et al. (2005)

Kim et al. (2005) numerically analysed the effect of an umbrella arch when employing forepole elements within a grout zone around the outside of the excavation, with parameters outlined in Table 1. The forepoles used in this study were 60.8 mm in diameter, 3 mm in thickness, and 12 m in length. They were installed with a S_{cfp} of 0.4 m and L_{fpo} of 6 m, and the α_{fpa} of the umbrella arch was 120°. The grout was simulated by multiplying the original ground deformation modulus by a factor of 2. Their results showed a greater impact on reducing the surface settlement when employing the umbrella arch system in weaker ground condition (see Fig. 5), compared to instances that did not include the system (but other support was installed). Through a retrogressive analysis, their results further indicated that a prediction of surface and tunnel crown settlement is possible. However, their results did not agree with the surface settlement reduction plot found in the empirically driven UASC (Oke et al., 2014a) as shown inFig. 5. A comparison of the results from Kim et al. (2005) with the surface reduction plot ofOke et al. (2014a) illustrates two important factors. Firstly, a parametric analysis with numerical models must be calibrated to a case study to bring any validity to the work and, secondly, due to the complexity of the umbrella arch system, it proves to be very challenging to be modelled numerically.

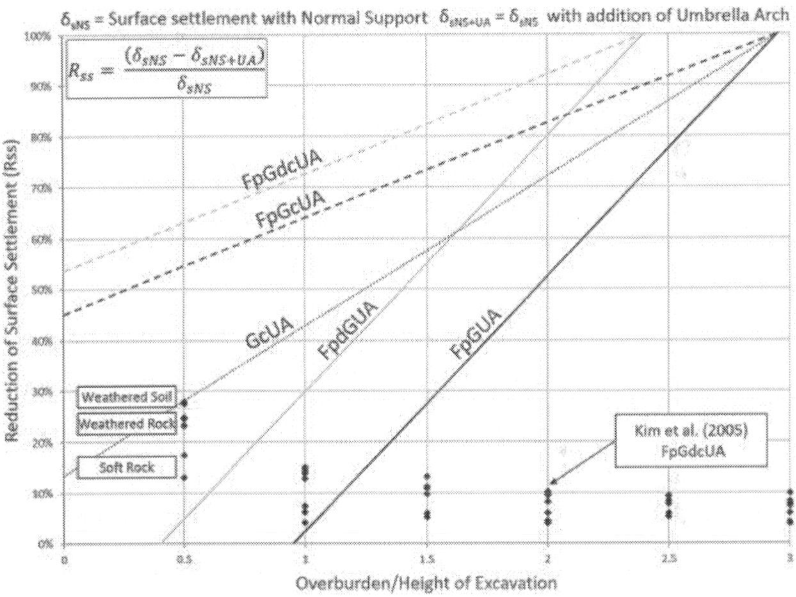

Figure 5. Surface settlement reduction plot from the UASC (Oke et al., 2014a). Black diamonds indicate the results found from Kim et al. (2005). Note: FpGUA = forepole grouted umbrella arch; FpdGUA = double forepole grouted umbrella arch; FpGcUA = forepole continuous grouted umbrella arch; FpGdcUA = forepole double continuous grouted umbrella arch; GcUA = continuous grouted umbrella arch.

NUMERICAL INVESTIGATION

The authors have conducted 2D and 3D parametric analyses in order to illustrate the challenges of modelling forepole elements numerically. In each respective case, analyses were conducted by industry standard programmes: 2D numerical analyses were carried out by employing Phase2 v7 (Rocscience, 2010) and v8 (Rocscience, 2013); and 3D analyses were carried out by employing FLAC3D v4 (Itasca, 2009). These analyses also took into account deep and shallow tunnel excavations. The deep excavation was based on the parameters used in Vlachopoulos (2009) andVlachopoulos et al. (2013) investigations of the Driskos tunnel of the Egnatia Odos highway in Greece. The parameters selected are from Section 4.3 of the Driskos tunnel where forepoles were employed in squeezing ground conditions of fractured flysch material. Selected ground parameters can be found in Table 2. The parameters provided the authors the opportunity to create a numerical model with previously verified input parameters (matched to average in-situ tunnel convergence). In an effort to further increase computational time, however, the simulation was simplified to a single circular tunnel excavation (full face), unless noted otherwise. An additional deep numerical model was also created based on a hypothetical squeezing case (hydrostatic condition at depth). The relevant general ground parameters of the generic squeezing model can be found in Table 2.

Table 2. Parameters used for numerical analysis. Materials were perfectly plastic, with no dilation. H_e is the height of excavation.

Case	Shape	D_t (or H_e) (m)	In-situ stress (or overburden)	Excavated material	Hoek-Brown parameters			Mohr-Coulomb parameters		Modulus of elasticity, E_{rm} (MPa)	Poisson's ratio
					m	s	a	c (kPa)	φ (°)		
Driskos tunnel (Vlachopoulos, 2009)	Circle	10	100 m	Flysch (GSI = 31)	0.66	0.000468	0.52	290	36	1442.1	0.25
Istanbul metro (Yasitli, 2013)	Horseshoe	6.8	10.75 m	Clay	–	–	–	20	33	38	0.33
Generic squeezing (Oke et al., 2014b)	Circle	10	3 MPa	Mudstone (GSI = 20)	0.345	0.0001	0.544	235	30.2	400	0.25

The shallow excavation runs performed were based on the in-situ surface settlement results and support design of the Istanbul metro, as published in Yasitli (2013) and Ocak (2008). Two different sections of the Istanbul metro had similar geological profiles and structural layouts. The major difference, however, was that one section employed an umbrella arch system with forepoles while the other section did not. This difference allowed the authors to validate the support system for both scenarios, and instilled confidence that the parameters used to simulate the forepole element were realistic. The parameters used to simulate the ground condition for the Istanbul tunnel can be found in Table 2.

2D numerical investigation

Literature concurred that 2D numerical analysis does not and cannot accurately simulate the response of forepole elements within an umbrella arch (Volkmann and Schubert, 2007 and Peila, 2013). In order to illustrate this inability, two types of simulations were conducted to analyse the forepole elements within an umbrella arch. The first was a homogenous model, suggested by Hoek (2001) as a crude approach (Fig. 6a). The second was an "as-built" model where the forepoles were simulated explicitly (Fig. 6b).Fig. 6a illustrates that the stresses appear to capture the radial aching effect around the outside of the homogenous region, but the model is not able to capture the local arching between the structural elements. Furthermore, the homogenous model is not able to capture the longitudinal stress transfer. For the "as-built" model, no additional resistance to the deformation is provided by the forepole as shown in Fig. 6. Fig. 6d also illustrates the convergence of the unsupported, homogenous, multiple "as-built" simulated models and multiple 3D numerical model results. The homogenous model is the only one that is able to capture the expected reduction in the crown convergence when other support members are installed. Analysis of this model has found that while the homogenous model might capture empirical trends of reduction of the crown convergence, this method does not, however, capture the true longitudinal mechanical response of the umbrella arch. This is because when the forepole elements are installed without other supports, there is no significant reduction in the crown displacement, as denoted by the 3D analysis results (squares in Fig. 6d). Furthermore, Peila (2013) stated that it is difficult or impossible to define the improved ground conditions. The authors agree with Volkmann et al. (2006) that the only acceptable application of the homogenous model exists when grout is continuously connected around the outside of the excavation (with or without steel reinforcement). Despite of considerations of this type of umbrella arch, it remains difficult or impossible to correctly select the accurate stress release action required (Peila, 2013) to capture the 3D tunnelling effect of the tunnel face, as explained in detail by Vlachopoulos and Diederichs (2014).

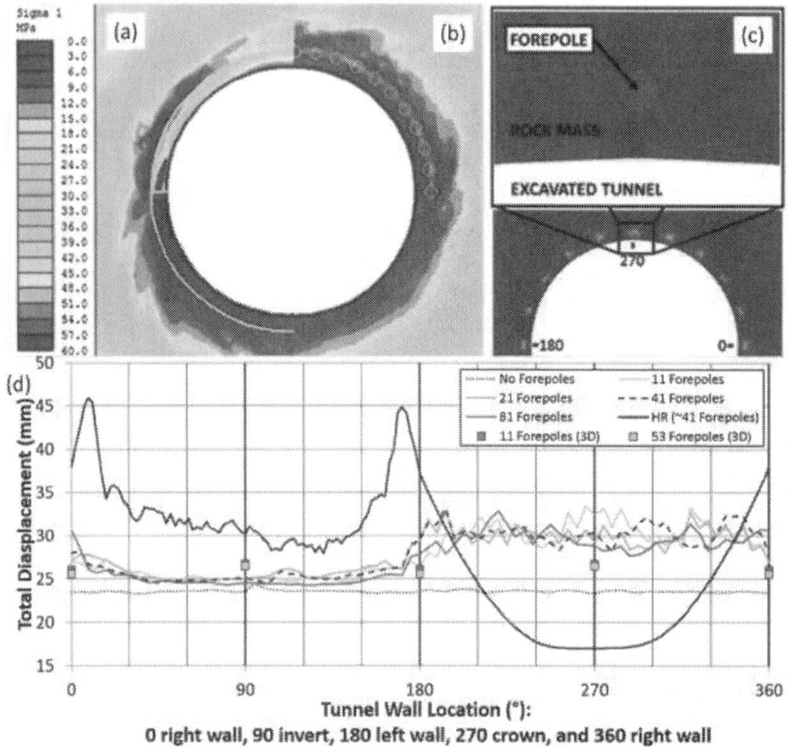

Figure 6. Phase2 analysis of principal stress relocation due to forepoles models as: (a) homogenous model, and (b) "as-built" model. (c) Illustration of layout of 11 forepoles around top half of excavation (2D analysis). (d) Results of parametric 2D and 3D analyses of quantity of forepoles versus tunnel convergence.

3D numerical investigation

As previously explained, the authors conducted parametric analyses based on two different case studies to aid in the design and understanding of the forepole structural element. Both case studies have been simplified to single excavations with a constant excavation profile (no niche profile). The authors understood that such a simplification would, in turn, change the response of the Driskos tunnel excavation; however, the material and structural properties were selected from a previously calibrated numerical model. The changes of excavation process and tunnel profile would have a minimal impact on the already calibrated input parameters. Further simplifications to the numerical models will be stated in the following sections. Still, the authors conducted a parametric analysis on the interaction parameters for the forepole structural elements in order to fully understand their influence before any parametric analysis took place. The numerical mesh and boundary conditions for this analysis of the Driskos tunnel can be found in Fig. 7a–c.

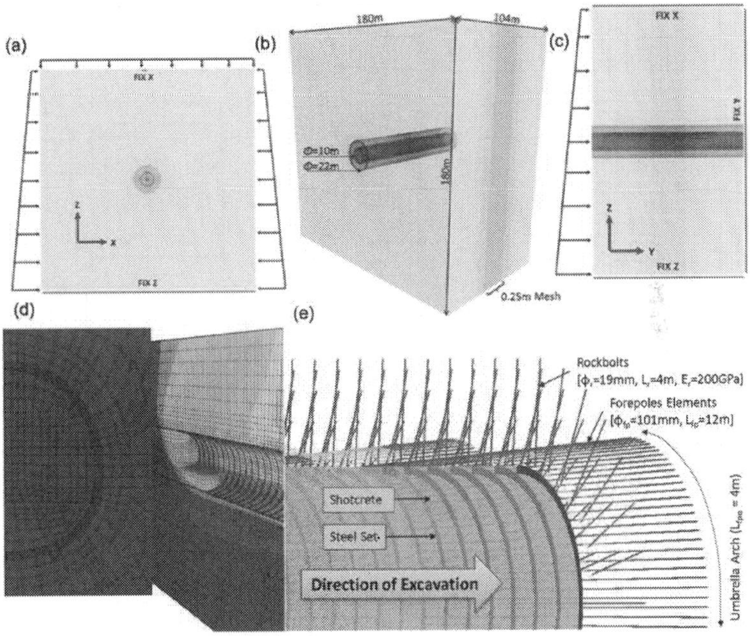

Figure 7. 3D numerical models used within this paper. (a)–(c) Driskos tunnel; (d) Istanbul metro; and (e) Generic numerical model.

The model set-up for the Driskos tunnel has a boundary width of $9D_t$ from the centre of the tunnel axis, and a longitudinal minimum boundary width of $4.5D_t$ (Fig. 7b). The boundaries are fixed at the bottom and at the entrance plane of the excavation in the normal direction. The top middle strip of the model is also fixed in the direction parallel to the tunnel axis (Fig. 7a and c). Stresses are applied on other boundaries to simulate gravity loads (Fig. 7a and c). The mesh was generated with a finer mesh (0.25 m) at the centre expanding to the peripheries. This primary numerical investigation of the Driskos tunnel simulated the forepole elements solely (no other support was simulated) in order to capture the independent influence of the support element, unless noted otherwise.

The numerical mesh of the Istanbul metro (Yasitli, 2013) can be found in Fig. 7d. It had similar model sizes and boundary conditions as the Driskos case except that a pressure of 100 kPa was applied to the surface (top) boundary to simulate the influence of building and traffic (Yasitli,

2013). The calibration of the Istanbul metro case was based on in-situ surface settlement results, which was also simplified to a single bore excavation. The simplification of the single bore excavation would cause an error within the numerical model, as the second excavation would influence the first one. To better understand this simplification, a simple 2D analysis was conducted. The authors utilised Phase2 v8 (Rocscience, 2013) to verify the impact of the twin tunnel excavation. A percent difference of 10% was found between the single excavation and double excavation when the umbrella arch method was not employed, as illustrated in Oke et al. (2013a). The calibration of the Istanbul tunnel and further parametric analysis will be explained in greater detail in the following section

Numerical investigation of forepole element

FLAC3D v4 (Itasca, 2009) possesses three different types of structural elements which can all be employed to simulate the individual forepole elements. These existing structural elements are CableSel, BeamSel, and PileSel (Fig. 8a). The CableSel element is capable of bearing axial loading only, and is not able to capture the longitudinal bending/stress transfer response of a forepole element. The BeamSel element is capable of taking on both axial and bending forces; essentially, the PileSel is a BeamSel element with the additional “rock-bolt logic”. The “rock-bolt logic” allows for the ability of the support element to account for changes in confining stress around the reinforcement, strain-softening behaviour of the material between the pile and the grid, and tensile rupture of the pile (Itasca, 2009). The slider constitutive model follows a Mohr-Coulomb failure criterion and the spring is defined by its stiffness parameter (Fig. 8c). The authors have found that the PileSel is the most suitable for the simulation of the forepole element, in agreement with other authors (Broch et al., 2006, Volkmann and Schubert, 2006a and Vlachopoulos and Diederichs, 2014) who have studied numerically the umbrella arch systems.

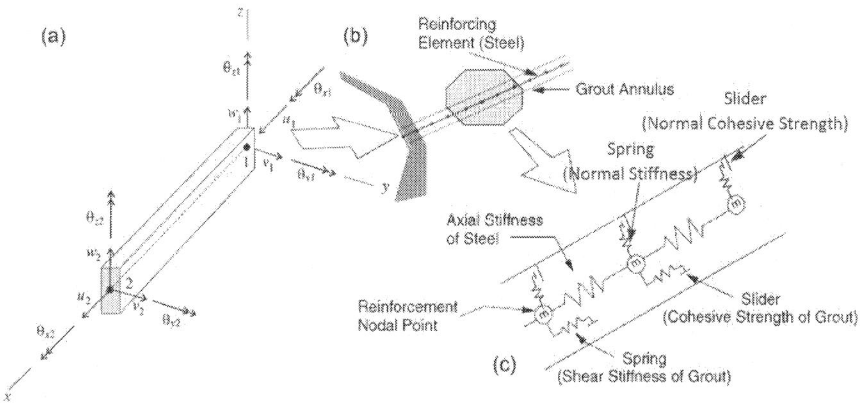

Figure 8. Illustration of the PileSel element in FLAC3D: (a) beam-column element; (b) nodal division; and (c) interaction parameters. Illustration is modified after Itasca (2009).

The concern that the PileSel element presents lies in the requirement that exact values of the interaction parameters must be selected (especially that of the stiffness parameter,Fig. 8c). Itasca (2009) suggested that these interaction parameters should be obtained from laboratory tests, yet, if laboratory testing of this nature is not possible, as is the case in most investigations, the stiffness parameter can be approximated by the method introduced by St. John and Van Dillen (1983). A simplified version of the St. John and Van Dillen (1983) equation (Eq. (1)) has been established to provide a reasonable calculation for the stiffness parameter, according to Itasca (2009). In addition, this simplification has a one-tenth factor which helps to account for the relative displacement that occurs between the structural element and the borehole surface (the annulus), as illustrated in Fig. 8b. The material properties regarding grout are not always provided, as is the cases of the Istanbul metro from Yasitli (2013) and the Driskos tunnel from Vlachopoulos (2009). Therefore, Itasca (2009) suggested as a general rule that the stiffness parameter be set to ten times the equivalent stiffness of the stiffest neighbouring zone (Eq. (2), with *m* value of 1). The stiffest neighbouring zone will always be the forepole element.

$$k = \frac{2\pi G}{10 \ln\left(1 + \frac{2t_g}{\phi_{fp}}\right)} \qquad (1)$$

$$K = m\ 10\ E_f \qquad (2)$$

where G is the shear modulus of surrounding material (usually grout), t_a is the thickness of the annulus, m is the stiffness multiplier, and E_f is the modulus of elasticity of the forepole element.

Identification of the correct stiffness parameter is essential. If the selected stiffness parameter is too low, the deformation of rock mass is beyond that of the support element, without capturing the true interactions. Conversely, if the stiffness parameter is too great, numerical instability is possible. Numerical instability is the result of the failure criteria of the interaction connection (slider) constantly resulting in failure with any slight movement of the numerical mesh around the structural element. An example of this sensitivity can be found in Fig. 9. Fig. 9 shows the displacement profile of the complete forpole element with 2 m having been excavated at the front end of the support element. The ensuing results are from adjustments of the interaction parameter by a factor of 10 for the generic squeezing numerical model, as previously described. The results illustrate that the greatest magnitude of deflection in this simulation was found when $m = 0$ and the cohesion, c, a parameter in the normal interaction direction was set to zero. When the value of m increased or decreased from $m = 0$, the maximum deflection value decreased.

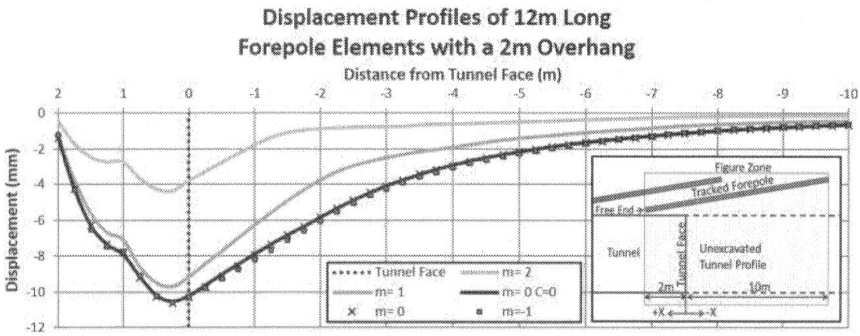

Figure 9. Parametric analysis of interaction stiffness parameter in order to determine the maximum displacement for a forepole element for the generic squeezing numerical model. Positive distance is within the excavated zone and negative distance is in the unexcavated zone of the tunnel profile. c is the cohesion,$k = 10mE_f$, where $E_f = 200$ GPa.

The interaction parameters for the forepole element of the Driskos tunnel were investigated in an effort to illustrate the sensitivity of each parameter, as shown inTable 3. The base model employed the parameters suggested by Itasca (2009) and the lowest values (expect practically zero values) were found in the literature (Trinh, 2006). The lower values of the sensitivity analysis were based on a zero or practically zero value and the higher values were based on the highest values found in the literature

(Vlachopoulos (2009) for FLAC3D and Funatsu et al. (2008) for particle flow code (PFC) (Itasca, 2002)) or a multiplication of the base value. An illustration of the interaction parameters between the pile elements and the numerical mesh can be found in Fig. 8c.

Table 3. Interaction parameters for the sensitivity analysis of the PileSel interaction parameters with the associated analysis runs.

Interaction parameters							
FLAC3D code	cs_scoh	cs_sfric	cs_sk	cs_ncoh	cs_nfric	cs_nk	cs_ngap
Symbol	c_s	φ_s	k_s	c_n	φ_n	k_n	g
Units	[F/L]	(degrees)	[F/L^2]	[F/L]	(degrees)	[F/L^2]	On/off
Base	[a]$c\phi_{fp}\pi$	[a]φ	[c]1.0×10^7	[a]$c\phi_{fp}\pi$	[a]φ	[c]1.0×10^9	Off
High value	$2c\phi_{fp}\pi$	2φ	[b]1.0×10^{11}	$2c\phi_{fp}\pi$	2φ	[d]1.0×10^{12}	On
Low value	0	0	0	[a]0	0	0	
High run#	Run 1	Run 3	Run 5	Run 7	Run 9	Run 11	Run 13
Low run#	Run 2	Run 4	Run 6	Run 8	Run 10	Run 12	

Note: c_s is the cohesion in shear direction, φ_s is the angle of friction in shear direction, k_s is the stiffness in shear direction, c_n is the cohesion in normal direction, φ_n is the angle of friction in normal direction, k_n is the stiffness in normal direction, "g" represents the gap, c is the cohesion of ground material, ϕ_{fp} is the outside diameter of the forepole element, and φ is the friction angle (Trinh, 2006, Funatsu et al., 2008,Itasca Consulting Group Inc, 2009 and Vlachopoulos, 2009).

The results of this sensitivity analysis are illustrated in Fig. 10 and Fig. 11. Fig. 10displays the results of the sensitivity analysis of the tunnel convergence recorded at the tunnel face and at the distance three times of the tunnel diameter away from the tunnel face ($3D_t$). The parameters which show an influence on the tunnel convergence in Fig. 10are plotted in Fig. 11. Fig. 11 presents the different displacement profiles captured in the sensitivity analysis. The most notable analyses from Fig. 11 are Run 12 and Run 8. Run 12 has a practically zero value for the normal stiffness value of k_n. Due to such a low k_n, little or no stress interaction occurs between the supports of the deforming ground. The results from Run 12 suggest that the majority of mechanical response from the forepole element comes from the normal stress interaction. Run 8 illustrates that the interaction parameter failure occurs between the forepole and the tunnel face, allowing the ground material "flow" around the structural element. The results of this sensitivity analysis find that the parameter k_n will govern the deflection profile of the forepole element. Furthermore, the cohesion, c_s, and the shear stiffness, k_s, are the second and third most influential parameters, respectively.

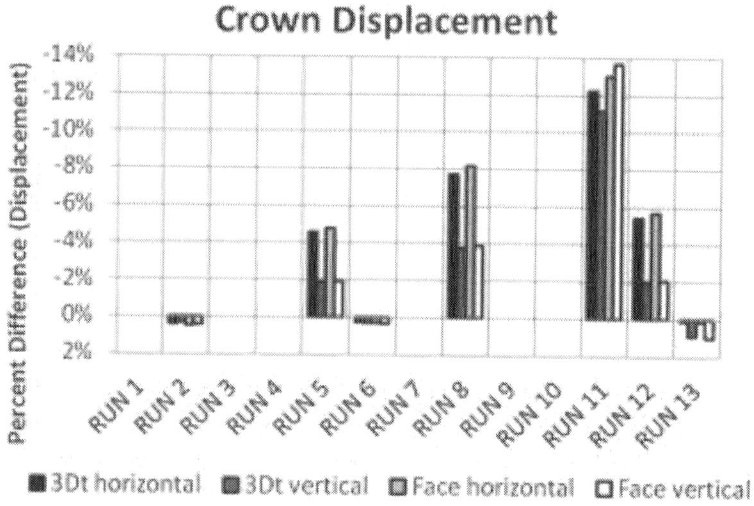

Figure 10. Percent difference of displacement with baseline interaction parameters values. $3D_t$ denotes the measurement of the convergence conducted at the distance three times of the tunnel diameter away from the tunnel face. Interaction parameters for each run can be found in Table 3 (modified from Oke et al. (2012)).

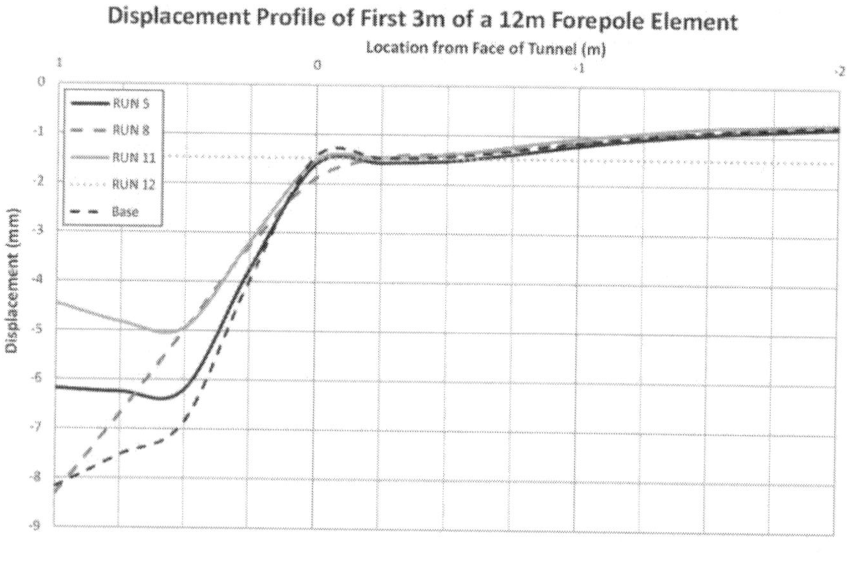

Figure 11. Displacement profiles of the first 3 m of a 12 m forepole element obtained by numerical analysis with 1 m of overhang. Runs that are too similar to base results are not shown in the chart. Interaction parameters for each run number can be found in Table 3.

The authors utilised the lessons gleaned from the interaction sensitivity analysis to calibrate the Istanbul metro case study. In order to calibrate the numerical model, the *k*value (normal and shear) was adjusted for all structural elements in situations involving tunnels built without forepole elements. This situation was captured within 5% of the in-situ data, which is in acceptable limits of possible error due to the aforementioned simplification of the single tunnel excavation. The calibration of the model with the forepole element required a more intensive process as illustrated in Fig. 12. Fig. 12represents the three step process carried out in order to find the greatest impact on reducing surface settlement due to the installation of the forepole element. First, the $k_s$and k_n multiplier, m, was increased (denoted by the black squares in Fig. 12) from 1 until it became evident that the minimum surface settlement was present. This minimum value occurred when the stiffness multiplier was 4. Next, the normal multiplier was kept constant at $m = 4$ while the k_s parameter multiplier was varied from 1 to 6, as denoted by the black diamonds in Fig. 12. As is illustrated, there was no change in the ground surface settlement until the k_s parameter multiplier became greater than 4. In the third of three steps, the shear multiplier was held constant at $m = 4$ while the k_n parameter multiplier varied from 2 to 5, as denoted by the black triangles in Fig. 12. The smallest amount of surface settlement was captured when the k_n multiplier was reduced to 3. Consideration of these results dictated that the forepole elements' shear and normal stiffness multipliers were set to 4 and 3, respectively, for analysis of the design parameters of the Istanbul metro. The calibrated numerical model with forepole elements was within 25% of the in-situ results, a difference of only 10 mm. Such a minor difference can be credited to an inaccurate capture of the interaction between the forepole element and other structural support members, as explained in the subsequent section.

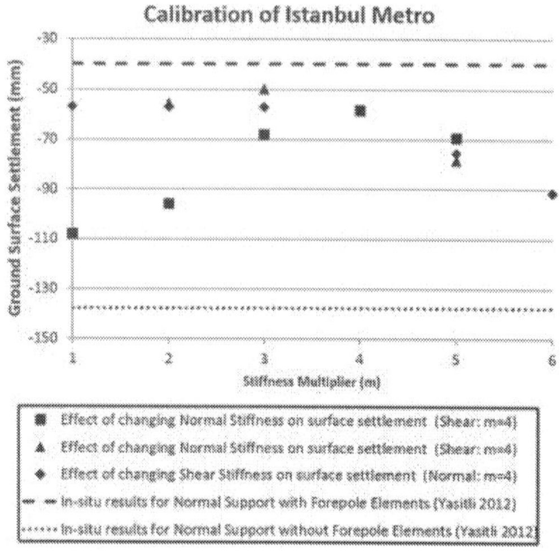

Figure 12. Numerical results of the Istanbul metro for calibration of the normal and shear stiffness interaction parameters.

Broch et al. (2006) found that a fixed connection between the spile elements and other supports could reduce the displacement ahead of the tunnel face by 80%, when compared to the un-fixed (free) numerical analysis. However, it remains the authors' opinion that a forepole embedded into shotcrete forms an elastic or plastic connection, not a fixed connection. To date, FLAC3D does not support multiple-layered interaction connection (Itasca, 2009), which makes an elastic/plastic connection difficult to be incorporated. In lieu of this, the authors felt that a free connection would most closely represent reality.

Design parameters investigation

As previously discussed, a parametric investigation in the Driskos tunnel and the Istanbul metro was conducted to illustrate the influence of each design parameter: S_{cfp}, α_{fpa}, the size/stiffness of the forepole element, α_{fp}, L_{fpo}, and the effect of other supports and geometry.

(1) Centre to centre spacing, S_{cfp}

Centre to centre spacing, S_{cfp}, of the forepole element of an umbrella arch was investigated by increasing the number of forepole elements that spanned a coverage angle of 180° from 3 to 157. Fig. 13a illustrates the results of this analysis. An increase in the number of forcpolcs from 3 to 53 yields a displacement at the free end of the forepole of 18% difference (denoted by the diamond markers within the embedded chart ofFig. 13a). Despite of this, there remains only a 5% difference of displacement when the number of foreploes was increased from 53 to 157 (denoted by the square markers within the embedded chart of Fig. 13a). Therefore, the addition of more forepoles in an effort to control deformation beyond this threshold is, perhaps, not significantly advantageous when balancing material cost and operational cost (as noted previously in the investigation of spacing by Volkmann and Schubert (2006a)). It is important to note that the mesh size of the numerical model was 0.25 m at the tunnel boundary, which is nearly equal to S_{cfp} of the 63 forepoles. Thus, after the number of forepoles is greater than 53 (~0.3 m spacing), minimal impact is induced by the mesh size of the numerical mode. Furthermore, as the numerical model used was a continuous model, failure that would most likely occur between elements with larger spacing was not captured, rendering the results of this numerical analysis applicable only for stable (at the tunnel face) squeezing ground conditions, such as the Driskos case.

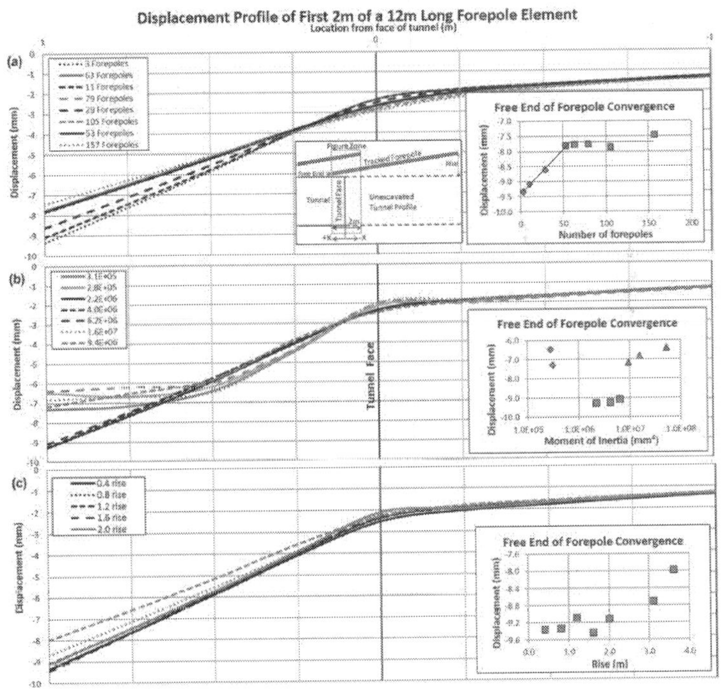

Figure 13. Displacement profiles of the first 2 m of a forepole element based on 3D numerical analysis with the Driskos tunnel parameters. The plot illustrates the values of displacement of the free end of the forepole with varying number of forepoles: (a) effect of number of forepoles; (b) effect of forepole size; and (c) effect of angle of installation.

S_{cfp} was further investigated by employing the numerical model based on the Istanbul metro ($\alpha_{fpa} \approx 170°$, $S_{cfp} = 40$ cm, $\phi_{fp} = 114$ mm). A numerical analysis was carried out to investigate the effect of varying centre to centre spacing from 26 cm to 50 cm while simultaneously changing the coverage angle from 90° to 200°. Twenty four cases were analysed to allow for a higher resolution of natural neighbour interpolation of data points with the ranges examined. The results of these analyses (see Fig. 14) illustrate a slight difference in reduction of surface settlement with the effect of spacing in comparison to the coverage angle. It is found that the coverage angle has a greater influence than the spacing of the forepole element on the global response of the system. It is also apparent from raw data, however, that the 40 cm spacing is capable of controlling settlement more effectively than spacing with an equivalent coverage angle of up to 160°, indicating an optimum spacing value. Once again, however, this analysis was unsuccessful at capturing the local failure between the support elements (due to mesh size), and the interaction parameters were only calibrated to a spacing of 40 cm. According to these results, the stress redistribution caused by installation of the forepole elements were captured, as illustrated in Fig. 15. Fig. 15 displays the final displacement of three different points around the tunnel cavity for varying coverage angles while maintaining constant spacing (40 cm). The results find that the stresses are first redistributed at the side walls, causing great deformations. As the coverage angle increases, the stresses are transferred away from the excavation walls to below the excavation due to the support, decreasing the convergence of the side walls further, which explains why the coverage angle has a greater impact on the global response.

(2) Size/stiffness of the forepole element

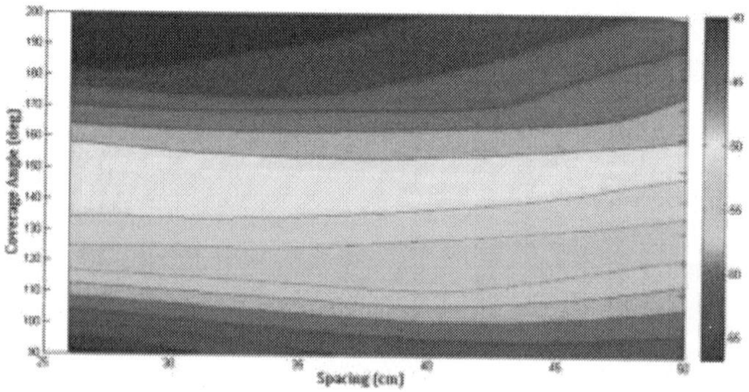

Figure 14. The effects of centre to centre spacing and coverage angle on surface settlement based on numerical models (parametric analyses of 24 cases) of the Istanbul metro.

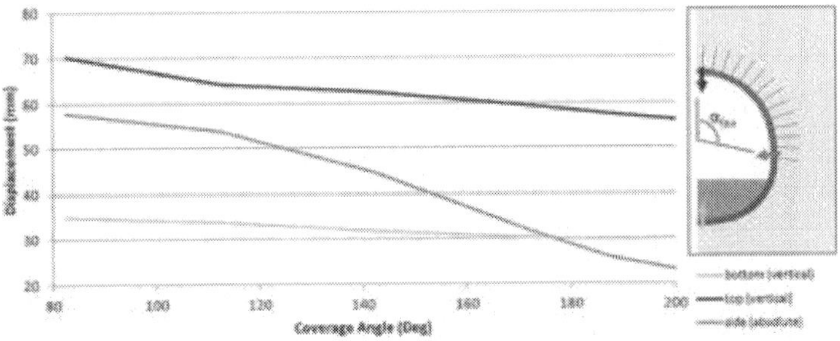

Figure 15. Effect of coverage angle on displacement for a 40 cm forepole spacing for the Istanbul metro analysis. Right embedded image displays a α_{fpa} of 200°, and the arrows illustrate the location of referenced displacement to the respected colour.

A parametric analysis based on the Driskos tunnel was also conducted on the size of the forepole element. The value of ϕ_{fp} was selected to capture the full range of acceptable forepole sizes. The smallest diameter was that of the largest standard size rebar used in Europe (50 mm). The range of forepole metal pipes used in the analysis was the values cited in the literature. Intermediate sizes were also evaluated in order to determine the effects of a constant t_{fp} (6.5 mm) while simultaneously altering the value of ϕ_{fp}, and a constant ϕ_{fp} (141 mm) while altering the value of t_{fp}. These numerical changes in the size of the forepole element, however, only affect the stresses being applied to the structural element (surface area) and the stiffness of the structural element. The purpose of these numerical runs tried to determine the existence of an optimal range of forepole size. Forepole elements require a minimisation of stress concentrations to ensure a large enough exposed perimeter to disperse stresses, but must be simultaneously small enough to move with the ground. The result of this analysis can be found in Fig. 13b. The optimum size for ϕ_{fp} was 101–141 mm, with a t_{fp} of 4–6.5 mm, which resulted in a moment of inertia range of $(2.2–6.2) \times 10^6$ mm^4 as denoted with black lines in Fig. 13b as well as the squares within the imbedded image.

(3) Angle of installation

The angle of installation (α_{fp}) was also investigated using the Driskos tunnel numerical model. The change of α_{fp} was adjusted by increasing the rise (offset length normal to the tunnel boundary) of the forepole element. The lowest rise was set to capture the extreme case whereby the forepoles were separated by 12 cm (to allow for the application of shotcrete and other relevant supports, and accessibility of equipment). The rise was increased slightly in order to capture the typical installation angles (3°–7°).

Further incremental runs were conducted until a rise of 3.6 m was achieved, in an effort to capture an installation angle greater than 15°. Such an analysis indicated that increasing the installation angle of forepole will decrease the displacement of the forepole. On average, the forepole also experienced 2.6 mm of displacement at the face. Conversely, a decrease in convergence was observed at the tunnel face (3.51%) and $3D_t$ from the face (2.82%) as the rise was increased from 0.4 m to 3.6 m. Within the forepole structural element itself, the numerical analysis captured a compression of 800 N for 0.4 m rise, which is in sharp contrast to a recorded tension of 4200 N for the 3.6 m rise. It can be postulated that with an increased angle, the forepole would act in a mode similar to that of a rockbolt rather than purely a forepole function. This conclusion is conducive with rockbolt design logic, usually most effective when the rockbolts are installed normal to the tunnel axis and take on tension, while forepoles bear axial compression force (Volkmann and Schubert, 2007).

The niche profile is defined by the angle α_{fp}, which creates a greater excavation opening with a larger angle. This analysis did not, however, simulate the niche profile and as such, did not capture the effect of the increasing excavation opening size with the increasing α_{fp}. From the authors' collective experiences, the minimal improvement to convergence for the numerical model does not outweigh the potential for further convergences based on the absent assumptions of the numerical model. Thus, the authors conclude from this analysis that the α_{fp} of a forepole element should not diverge significantly from the horizontal plane unless structural analysis is conducted to ensure that the support is analysed in accordance with various modes of failure. The results of this analysis can be found in Fig. 13c.

(4) Overlap of umbrella arch

An investigation was carried out to determine the effect of overlap of the umbrella arch,L_{fpo}, design for the Istanbul metro ($L_{fpo} = 3$ m, $L_{fp} = 12$ m). This analysis, however, was modified to a full face excavation in order to remove the influence of staged excavation (i.e. top heading and bench) on the tunnel face and to promote more convergence at the tunnel face. Different overlaps of 0 m, 3 m, 4 m, 5 m and 6 m were selected for the analysis. Fig. 16 illustrates the improving trend of L_{fpo} when (a) the density of the face reinforcement has been kept consistent with the original design and (b) the face reinforcement is not altered from the original design (support only installed within top heading region). In both cases, the surface settlement of the numerical model converges when L_{fpo} is greater than the Rankine block failure distance (RFD) (i.e. the distance away from the face that the RFD passes), as illustrated in Shin et al., 2008 and Wang and Jia, 2008, and Volkmann and Schubert (2010). This finding indicates that, in order to optimise the effect of umbrella arch on surface settlement, forepoles should always be installed just past the failure zone distance, ahead of the tunnel face. Furthermore, it was found when the umbrella arch was not installed, there was only a 13% improvement of the surface settlement by increasing the face reinforcement from case (b) to case (a). Comparably, when the umbrella arch was installed (3 m of overlap only), there was a 58% improvement of surface settlement when the face reinforcement was increased from case (b) to case (a), as shown in Fig. 16. Such differences regarding surface settlement between the two face reinforcement patterns are caused by the redistribution of stresses at the invert of the tunnel, resulting in larger displacements at the bottom of the tunnel face. These results indicate the importance of well-designed face reinforcements and the requirement for further investigation of the impact of face reinforcement and umbrella arch interaction.

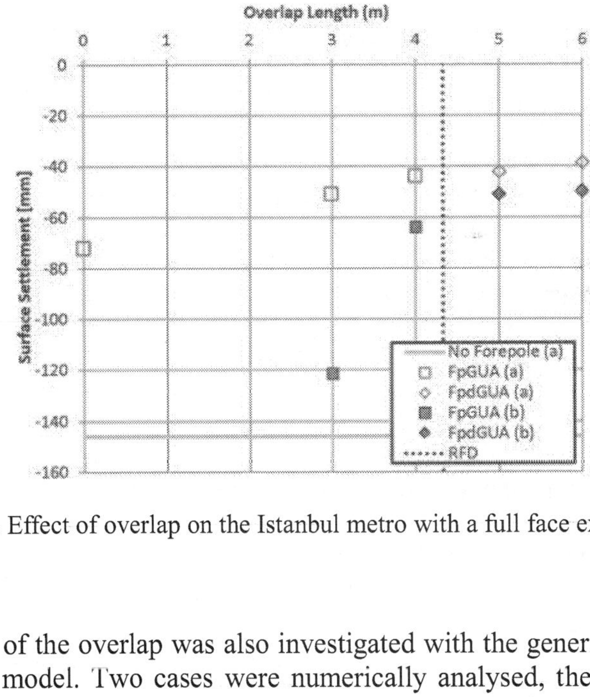

Figure 16. Effect of overlap on the Istanbul metro with a full face excavation.

The effect of the overlap was also investigated with the generic squeezing numerical model. Two cases were numerically analysed, the first with a 4 m overlap and the second without overlap. The deflection profiles of the two numerical models are plotted in Fig. 17where only 1 m of the forepole element is shown. The location of the free end of the forepole element was 2 m from the tunnel face. The analytical models proposed by Oke et al. (2014b) were calibrated to the two different cases by the least square analysis for both loading conditions defined by the longitudinal displacement profile (LDP) byVlachopoulos and Diederichs (2014) and the modified LDP by Oke et al. (2013b). It was therefore determined that by curve fitting of the deflection profile, a factor (0.5 and 0.4 for the LDP and the modified LDP, respectively) of the loading condition within the region of the overlap was required in order to match profiles. This is a promising result which captures the additional benefits of the overlap. However, additional investigation is required with regard to in-situ results in order to further quantify and validate this reduction of the loading condition. The analytical process is outside the scope of this paper, and will not be further discussed within this paper. Further explanation of this analytical process can be found in Oke et al. (2014b).

(5) Other support elements and geometry

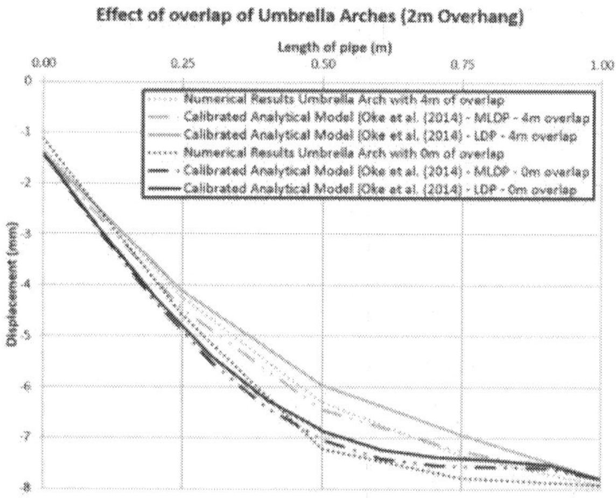

Figure 17. Results of numerical models based on the generic squeezing tunnel and semi-analytical models based on the proposed model of Oke et al. (2014b). The 4 m overlap case had a loading condition reduced by a factor of 0.5 and 0.4 for the LDP and modified LDP, respectively.

The impact of additional temporary support elements on the response of the umbrella arch support system has been found to be critical. The empirical evidence used to create the UASC (Oke et al., 2014a) shows that with increasing overburden, there should be a greater reduction in surface settlement. However, the numerical model of the Istanbul metro illustrates that changing the overburden by 2 m induced an increase of 19% of surface settlement, as denoted by the 5 solid black outlined squares in Fig. 18 (Oke et al., 2014c). This increase can be attributed to the 30% increase of the convergence of the excavation as the overburden increases and as the support system remains unchanged. Negative trends with respect to overburden were found when double forepole grouted umbrella arch (FpdGUA) was investigated, as denoted by the 5 solid black outlined triangles in Fig. 18. These results, as well as the results found in Kim et al. (2005), indicate that the design of the umbrella arch support requires the inclusion of some type of factor associated with the remaining support system employed. Furthermore, the results illustrate that the circle geometry creates less surface settlement (when the umbrella arch is not installed) of 22% and 46% for the 6.5 m and 6.0 m diameter tunnels, respectively, when compared to the as-built case for the Istanbul metro. When an umbrella arch was installed for the 6.0 m and 6.5 m diameter tunnels, the reduction of surface settlement was found to be 53% and 58%, respectively. Therefore, the results of circular tunnels found that the forepole grouted umbrella arch (FpGUA) had less influence on reducing the surface settlement than those installed for a horseshoe tunnel. Similar results were found when the FpdGUA was simulated in the Istanbul model, and 73% and 71% reduction of surface settlement occurred for the 6.5 m and 6.0 m diameter tunnels, respectively.

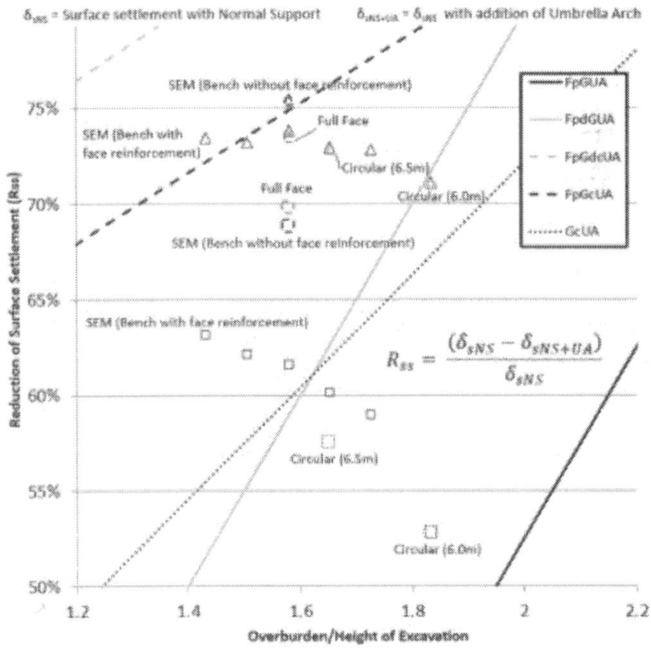

Figure 18. Numerical analysis results from the Istanbul metro plotted on the subsidence management plot from the UASC. Squares denote analysis conducted with a FpGUA, and triangles denote a FpdGUA. SEM = sequential excavation method.

OPTIMISATION METHODOLOGY AND VALIDATION

Both the literature and the numerical investigations provide findings which support the inclusion of numerical assessment for the spacing and overlap design parameters in the umbrella arch system, prior to analysis of the global response. Such numerical assessments are required as these design parameters are based on local failure mechanisms which remain difficult to be captured (and quantified) in a numerical analysis of the complete tunnel excavation. In the following sections, the authors will propose an optimisation methodology which employs numerical analysis to select design parameters. This optimisation will be conceptually validated using the worst squeezing case scenario captured at the Driskos tunnel at section 8 + 746. At this location, the FpGUA did not fail, so it is eligible as validation for the design methodology. The worst section (Chainage 8 + 746), however, included displacements far greater than the average parameters used in previous analysis. Therefore, further investigation is required and will be presented in the following section.

Driskos twin tunnel construction project

As previously stated, the preliminary properties conducted within this paper and used for the Driskos tunnel analyses, were found prior to the tunnel excavation. Upon excavation, it was discovered that at the selected location the squeezing of ground material was far greater than anticipated (Vlachopoulos, 2009). The maximum tunnel closure recorded during the top heading was found to be 210 mm (Egnatia Odos, 2001), along with primary-support failures along a stretch of the left bore (Chainage 8 + 500 to 8 + 800) (Grasso et al., 2005). From Chainage 8 + 657 to 8 + 746, monitoring data were collected and presented by Vlachopoulos (2009), as shown in Fig. 19. The calibration process was carried out from the worst case of convergence from in-situ data, Chainage 8 + 746. Despite of the data existing as an isolated condition, the forepole elements did not fail, rendering the design recommendation functional for these conditions. While the preliminary properties successfully captured the trend of most of the in-situ data at Chainages 8 + 657, 8 + 697, and 8 + 724, as illustrated in Fig. 19, they underestimated the displacement of the isolated condition at Chainage 8 + 746, despite of the numerical model not having included any support elements. A back analysis conducted by Marinos et al. (2006) on a similar rock mass found that the σ_{ci} value was greatly overestimated for this isolated case, and must be reduced from 26.25 MPa to 5–6 MPa for the given Driskos overburden. The reduction of rock mass parameters can be found in Table 4. This reduction allowed for displacements of an unsupported model analysis far greater than that of the in-situ supported model, as shown in Fig. 19. The authors have conducted a calibrated as-built numerical model simulation of the Driskos tunnel section 8 + 746, as shown in Fig. 20. Only the spacing and overlap design parameters, however, are altered within this investigation; complete optimisation for the other design parameters (ϕ_{fp} = 101 mm, t_{fp} = 6.3 mm, L_{fp} = 12 m, α_{fpa} = 160°, and α_{fp} = 5.73°) will not be conducted. It is important to note that the zero reading for the in-situ data was not taken till at least 10 m back from the tunnel face (as illustrated in Fig. 19), making it difficult to calibrate the numerical model to any displacement near the tunnel face.

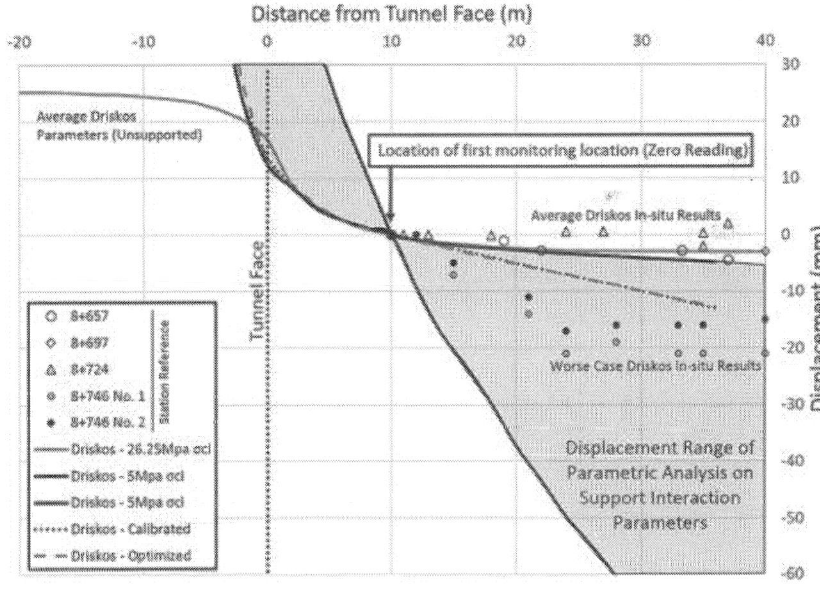

Figure 19. Comparison of in-situ data to numerical results of unsupported model. Hollow markers denote typical recorded data and filled markers denote the isolated squeezing condition. The grey shaded region represents the range of possible solutions of supported model based on numerical parametric analysis of the isolated ground material found in Table 3

Table 4.Parameters used for numerical analysis of the as-built Driskos tunnel at Chainage 8 + 746 (Marinos et al., 2006 and Vlachopoulos, 2009).

Shape			Height of excavation, H_e (m) (top/total)		In-situ stress (overburden) (m)	Excavated material
Horseshoe			8.78/11.70		106	Flysch ($GSI = 22$)
Hoek-Brown parameters			Mohr-Coulomb parameters		Modulus of elasticity (MPa)	Poisson's ratio
m	s	a	c (kPa)	φ (°)		
0.4	0.00	0.5			681.1	0.25
78	0172	38	118	21.3		

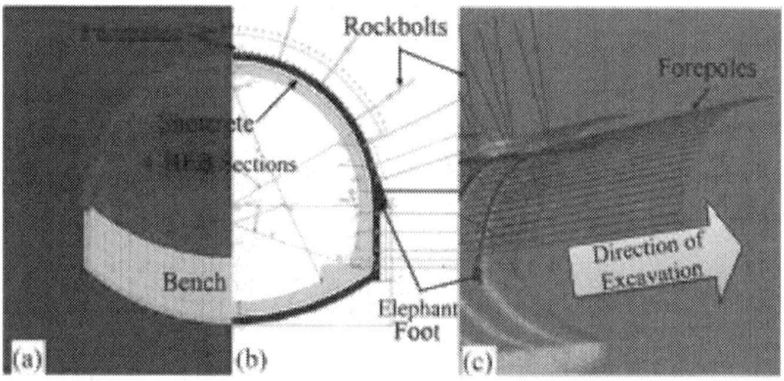

Figure 20. Illustration of the numerical model and support layout of the Driskos tunnel. (a) Cross section image of the numerical model of the Driskos tunnel. (b) Support layout used at the Driskos tunnel project (Egnatia Odos, 1998). (c) Oblique image of the as-built design of Driskos tunnel with visual support element used for analysis. Only top heading is excavated.

Spacing assessment

As previously mentioned, an investigation is required to find the maximum spacing for the forepole elements within the umbrella arch arrangement. Otani et al. (2008) proposed an experimental design spacing based on the diameter of the forepole and the friction angle of ground material. The result of this experimental design criterion theoretically maximised the height of the failure region to the radius of the forepole. However, this design did not take into consideration the effect of cohesion. The assessment from Doi et al. (2009) found that the friction angle had a minimal impact on the failure compared to the cohesion. In order to assess the possible maximum spacing, Doi et al. (2009)performed a trap door test using a discrete particle element model, PFC (Fig. 3 and Fig. 21a). They employed the forepole elements as rigid (fixed) particles and sequentially removed the boundary condition between and underneath the elements. As previously stated, Doi et al. (2009) captured the effect of cohesion and friction angle, which was then related to the local failure height between the forepole elements. This height, however, was not associated with any design standard to replace the work ofOtani et al. (2008).

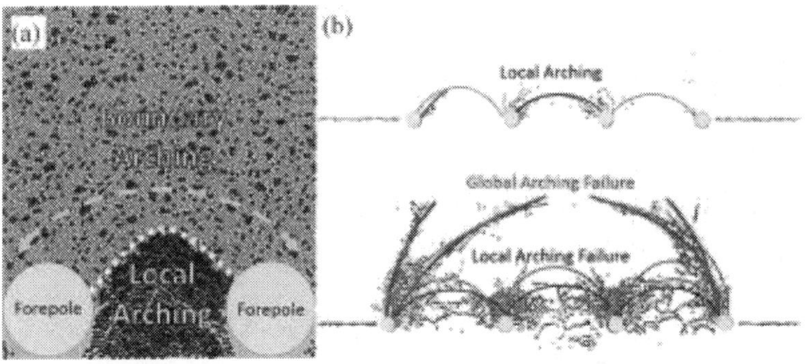

Figure 21. PFC numerical modelling of forepole spacing. (a) Illustration of boundary arching effect of the ground caused by the imposed boundary conditions, modified after Doi et al. (2009). (b) Illustration of local arching and failures, modified after Stockl (2002).

Based on the work of Stockl (2002), shown in Fig. 21b, Volkmann and Schubert (2007)indicated that the required spacing was designed to maintain local arching. In this event, the height of failure should not be an indication of the spacing design parameter, but instead be based on the forepole spacing ability to prevent ravelling type failure between forepole elements. This type of failure mechanism could not be captured in Doi et al. (2009) as the boundary conditions were not located in the centre of the forepole element.Doi et al. (2009) imposed a mirror on the outside edge of the model to simulate forepole elements with irregular spacing (i.e. side by side, space, side by side). This irregular spacing further resulted in a boundary arching effect, as shown in Fig. 21a.

Stockl (2002) performed a small scale physical experiment to illustrate two failure modes: local and global (Fig. 22). He illustrated that both of these failure modes could be captured within a comparable PFC numerical analysis (Fig. 21a). The authors decided to therefore illustrate this method through the creation of a similar type of analysis geared toward discovery of the optimum forepole spacing. Two such methods are illustrated within this paper, the first is a simplified model (continuum) while the second is an advanced model, such as the one proposed by Doi et al. (2009) and Stockl (2002).

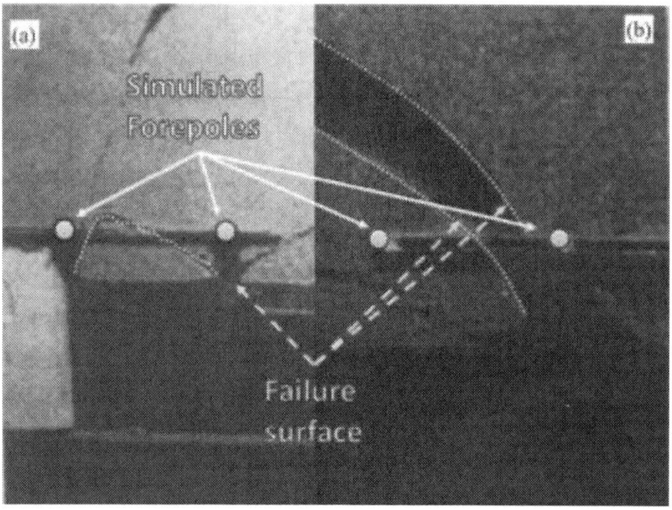

Figure 22. Physical testing results of Stockl (2002). (a) Local failure between forepole elements. (b) Global arching failure across forepole elements.

Simplified model

The simplified model will be illustrated by employing the Phase2 modelling programme, based on a homogenous material, for the ground and element zones to represent the grout and forepole elements. The boundary conditions and mesh of the simplified model are illustrated in the top portion of Fig. 23, which also illustrates the resulting options for loading conditions (field stress and gravity). The bottom boundary stress condition is lowered from in-situ to zero. When the loading conditions were compared with the calibrated numerical results of Stockl (2002), the gravity driven stress condition represents reality more accurately due to the shear failure developed above the forepole elements. The overall results, however, tend to develop a comparable tension failure region once the bottom pressure boundary is completely removed. Furthermore, the Phase2 analysis is based on a continuum model, and the ground material cannot fall between the forepole elements, as illustrated in the model by Doi et al. (2009). Based on the rigidity of the forepole elements and the imposed boundary conditions, a local arching effect will be consistently achieved (within realistic spacing parameters). The model, however, is able to capture the tension failure region, which, in theory, would has fallen out during the excavation.

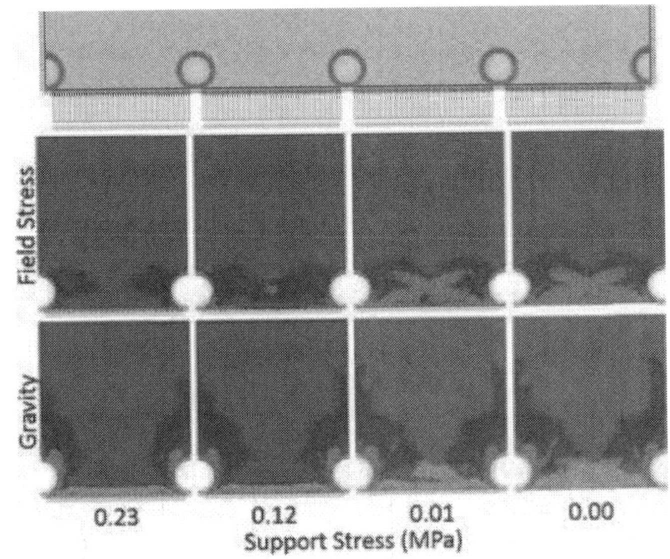

Figure 23. Illustration of the simplified spacing assessment for a 114 mm diameter forepole elements with 50 cm spacing. Top – mesh and boundary conditions of numerical model; Middle – field stress driven failure region propagation, and Bottom – gravity driven failure region.

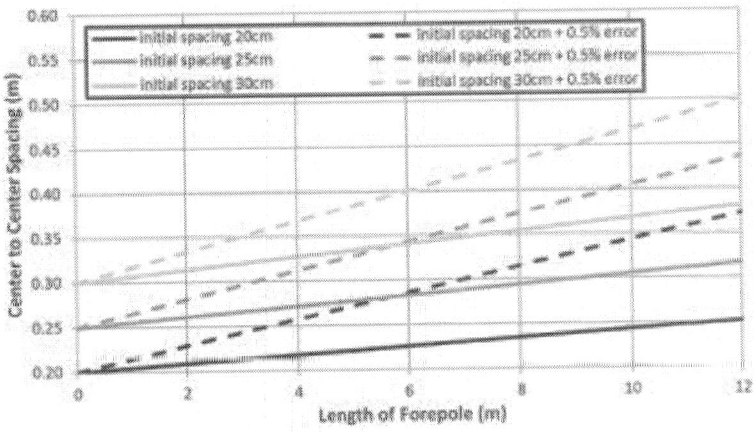

Figure 24. Spacing assessment based on a 12 m length of forepole element, with a α_{fp} of 5.73°.

An analysis was conducted on the spacing based on the average Driskos tunnel parameters. The tension failure region was found to be greater than the radius of the forepole elements when the spacing was increased to 35 cm. This result agrees with the as-built design of an initial spacing of 30 cm for the forepole elements. The spacing of the forepole element, however, is not constant due to the angle of installation of the forepole elements. The forepoles at the Driskos tunnel project were installed at an angle of 5.73° which would result in a spacing of 36 cm after 8 m of excavation (the location of the next installation of forepole elements). An even greater spacing allowance could exist (44 cm) if the calculation included the 0.5% average deviation of drilling (Mager and Mocivnik, 2000). Therefore, as an alternative to Otani et al. (2008), the authors hypothesise that a more appropriate design standard would designate that the maximum design spacing of the forepole element be based on the failure region height, equalling the diameter of the forepole element, as was found to be the average in the Driskos tunnel case.

To further validate this failure region height, an analysis was carried out using the new properties for section 8 + 746 of the Driskos tunnel. It was concluded that the spacing could not be determined without consideration of the other support members and the effect of the tunnel face. The support pressure (bottom boundary condition) was reduced to the mobilised support pressure (0.32 MPa) based on convergence-confinement theory (Carranze-Torres, 2004, Hoek, 2007 and Vlachopoulos and Diederichs, 2014). It was found through the simplification model that a local failure between the forepole elements would occur between a spacing of 40–45 cm, as shown in Fig. 25. This value is in agreement with the theoretical maximum spacing that occurred within section 8 + 746 (44 cm) due to a 0.5% installation deviation and installation angle of 5.73°.

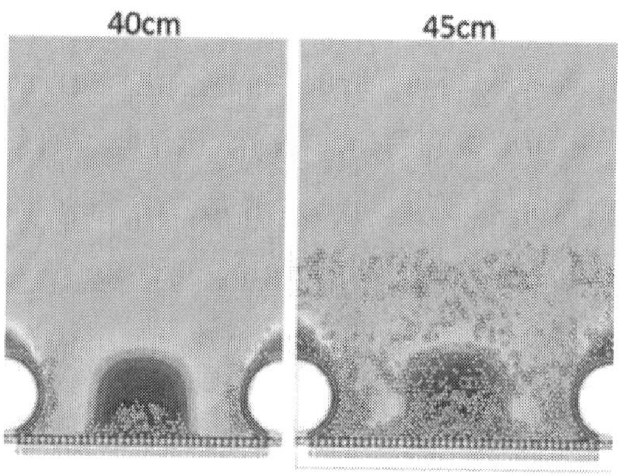

Figure 25. Simplified analysis of the new Driskos tunnel spacing assessment, with an internal pressure of 0.32 MPa. Contour of displacement: Red = minimum (~0 mm); Blue = maximum; White dots denote tension failure. Shear failure is not shown.

Advanced models

The previous results may prove successful for homogenous materials, but may not be applicable for more complex material models where stability of the ground structure may be the governing factor (gravity driven failures). It is important to note that all of the further analysis of the spacing assessment will be performed with 50 cm of spacing in order to promote the various failure modes that potentially could be exhibited by advanced numerical models.

While advanced models are able to capture the structural failure mechanism, the models do require calibration of the material parameters. Such calibration can be performed with the ground material simulated as a particle or discrete element model, or continuum (such as Phase2) model with simulated joints or Voronoi grains. Calibration is required to ensure proper interaction between particles or joints, as shown in Potyondy and Cundall (2004) for PFC models.

As an alternative to discrete particle codes, Fig. 26 provides an illustration of a Voronoi model and displays two of the four different automatic Voronoi mesh generations built into Phase2. The interaction parameters and the average size of the Voronoi model were held constant for the two different models, while the coefficient of uniformity was altered. It is apparent that the coefficient of uniformity has an additional impact on the design of the optimum forepole spacing, and should be taken into consideration for design. This Voronoi process can also be easily simulated with discrete element models.

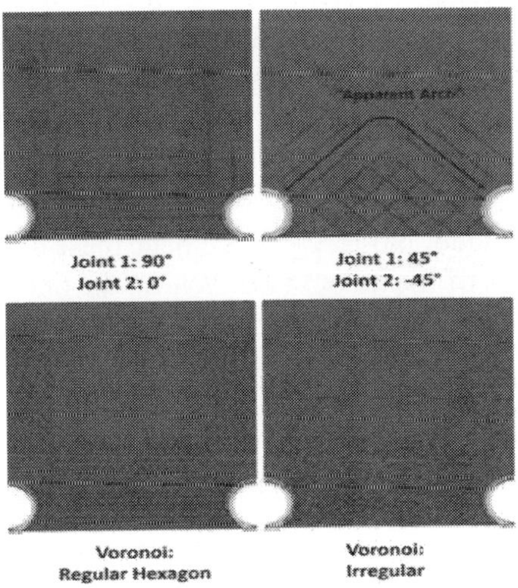

Figure 26. Illustration of the use of Voronoi and joint sets models to define the failure region for a 50 cm forepole spacing with gravity driven stress condition. The red lines indicate the failure joint surface of the Voronoi mesh or joint surface.

When assessing the spacing of the forepole element, another aspect that must be taken into consideration during the design process is the joint network of the ground material. An illustration of the impact of the joints on the response of the spacing can also be found in Fig. 26. The top left side of Fig. 26 illustrates a hypothetical condition in which joints are offset by 90° but remain parallel and perpendicular to the tunnel boundary surface, respectively. The top right side of Fig. 26 illustrates a hypothetical condition where joints remain offset by 90° but are 45° from the tunnel boundary surface. It is apparent that if the numerical model was not constrained within a continuum approach, a triangular shape block would have failed between the two forepole elements until the "apparent arch" was formed. Similarly, on the left side, a possible rectangular block would have failed between the two forepole elements. With regard to the failure limit, however, it is not explicitly clear where the rectangular block failed or that to the exact height the apparent arch would have propagated to. This joint model process, which was simulated in Phase2 with an explicit joint boundary, could be easily simulated with UDEC, but is not illustrated within this paper.

Though relatively simple, these advanced models can provide great insight into the mechanical interaction which exists between the individual forepole elements and the ground. Due to the demands on time, this type of assessment is not economically feasible (time requirement) within a complete 3D numerical analysis of the tunnel excavation. All other parameters, however, are capable of assessment in terms of the global response of the umbrella arch system, such as the overlap of the forepole elements.

Overlap assessment

As previously illustrated, the forepole element must be embedded past the disturbed zone ahead of the tunnel face. Within the numerical investigation of the Istanbul metro, the Rankine active failure block was a suitable approximation of the required overlap of the forepole elements of the umbrella arch. The approximation, however, does not include the impact of other face stabilising techniques. Limit equilibrium analysis could be taken into account for the other stabilising techniques. This type of analysis is outside the scope of this paper, but will be investigated in future research. Another approach available for assessment of the overlap requirement is the axisymmetric analysis as it requires merely hours for capturing, as opposed to days for a complete 3D numerical analysis. This axisymmetric analysis must be used with caution, however, as it is only truly applicable for installation of the forepole element in squeezing ground condition.

Axisymmetric analysis can be performed to illustrate the required overlap by assessing the distance from the tunnel face, along the tunnel boundary, to the outer limits of shear strain failure or the extent of plastic failure. An illustration of the extent of plastic failure can be found in Fig. 27. The additional face stabilising techniques can be simulated within the numerical analysis to capture their effect on the reduction of the required overlap. Structural supports within Phase2 analysis, however, cannot be simulated within axisymmetric analysis (except liners). Therefore, simplifications and/or approximations must be taken into consideration to simulate the face stabilising techniques. To illustrate simplifications and/or approximations, the authors have conducted an assessment on face reinforcement (soil nails) simulated as internal pressure acting at the face and improved ground conditions (Fig. 27). As is illustrated from the numerical analysis results presented in Fig. 27, the support (shotcrete only) is capable of reducing the failure region ahead of the tunnel face. It is also apparent that face support (simulated by an applied pressure or improved ground condition) will further reduce the failure in the vicinity of the tunnel face. Therefore, before embarking on a time consuming 3D numerical analysis, a simple axisymmetric model can be substituted to find the failure region ahead of the tunnel face and to help define the overlap required for successive umbrella arches.

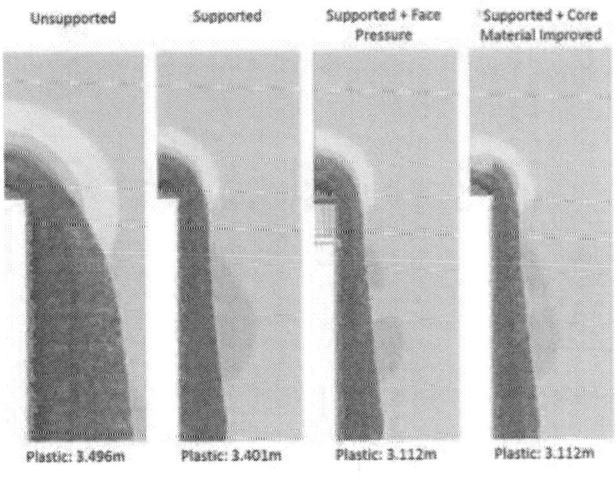

Figure 27. Illustration of the impact of different supports (and simulations of support) on the shear failure distance from the tunnel face along the excavation profile boundary.

The requirement for overlap for a squeezing ground condition is completely different from that for a subsidence driven condition. To ensure optimal use of the support in terms of reducing surface settlement, it is required that the next umbrella arch of support is installed while the embedded end of the current umbrella is in stable ground. In squeezing ground conditions, the plasticity zone is fair greater, and it remains impractical to install an overlap with this guideline. Furthermore, the rationale for installing forepoles under squeezing ground conditions is to transfer the stresses longitudinally away from the tunnel face which will, in turn, reduce degradation of the rock mass through confinement. Therefore, minimal embedment length, as opposed to subsidence, is required. From these considerations, the authors proposed that the selection of the overlap should be based on the maximum distances of the tension failure in front of the tunnel face. The required overlap, based on section 8 + 746, was found to be 1.488 m for the unsupported model and 2.412 m for the supported model, as shown in Fig. 28. The design for the overlap is also based on the pre-determined excavation step, which was 2 m. Therefore, the optimum overlap should be 2 m.

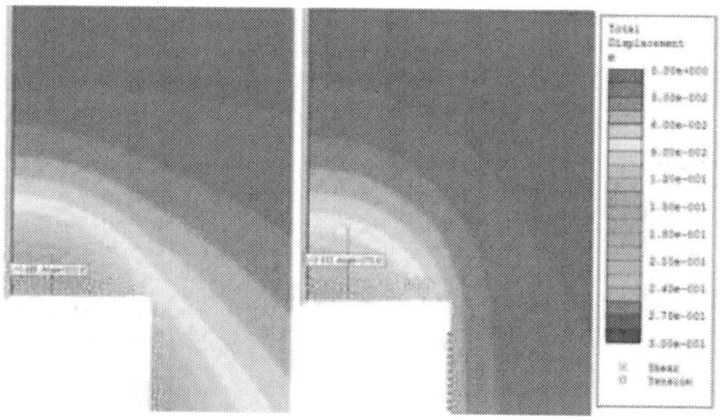

Figure 28. Tension failure of an axisymmetric analysis for the Driskos section 8 + 746. Left: unsupported. Right: supported (shotcrete only). Shear failure is shown within image.

Driskos tunnel design optimisation

As previously stated in Section 6.1, without changing/optimising the remaining design parameters, the result of the optimised initial spacing was found to be 25 cm, based on a maximum spacing of 40 cm with an overlap of 2 m for a 12 m long forepole element, as illustrated in Fig. 24. As it was shown in Fig. 19, a "calibrated" numerical model was found within the range of the parametric analysis of support interaction parameters. This calibrated model was created attempting to match the displacement trend from the worst case in-situ parameters. The exact profile was not able to be captured, which was attributed to simulating a homogenous material throughout the whole numerical model, as well as not simulating the niche or saw-tooth profile. However, the results are within reasonable limits for comparison of the optimised simulations. During the optimisation of the umbrella arch support, layout of the forepole elements with the simulated overall displacement profile did not change significantly, as shown in Fig. 19. The maximum resulting moment comparison, conversely, found that there was a 27.0% and −26.9% difference of positive and negative moment, respectively. This large difference of moment, however, was insignificant because the maximum resulting moment was over 100 orders of magnitude greater than forepole capacity found by Volkmann and Dolsak (2014). This finding is an indication either that the forepole elements stiffness was over designed or required to provide safe and effective excavation process for squeezing ground conditions. Furthermore, the optimised design resulted in an 8% reduction in tangential stress acting on the tunnel walls.

Furthermore, an evaluation of the economic impact of this design change was conducted. For each umbrella arch installed, 10 more forepoles would be employed to keep the same α_{fpa} when changing the initial spacing from 30 cm to 25 cm. This increase in forepole elements would require more time to install, slowing the excavation process. However, the decrease in the requirement for L_{fpo} over a 40 m stretch (before the next umbrella arch is installed) results in one fewer umbrella arch installation set-up when changing L_{fpo} from 4 m to 2 m. This optimised design results in one fewer umbrella arch installation set-up and 10 fewer forepole elements total employed over 40 m of excavation. Therefore this optimisation of design will decrease the time required to install by 10 fewer forepoles and one less set-up of the forepole jumbo for each 40 m of excavation, increasing the excavation rate.

SUMMARY AND CONCLUSIONS

This paper has investigated relevant concepts with respect to the employment of numerical techniques and analysis in support of the design of umbrella arch systems. An examination of 2D numerical analysis determined that such techniques were mechanistically incorrect in capturing the global response of such support systems, yet efficient in capturing localised failure between support elements and with respect to support-ground interaction; meanwhile, 3D analysis proved necessary for the capture of the global response of forepoles within an umbrella arch system. Through a detailed illustration of multiple, relevant sensitivity and parametric studies highlighted in the multiple sections of this paper, the importance of interaction parameters was emphasised, and procedures of recommended calibration processes were presented for both shallow and deep tunnelling excavation scenarios. The most sensitive parameters were determined to be k_n, k_s, and C_n. The individual influence of each respective design parameter on the global response was also presented.

The length of overlap, L_{fpo}, was found to be related to the investigated failure region ahead of the tunnel face. For subsidence driven (shallow) designs, L_{fpo} must be at a distance past the plastic failure region. For squeezing ground, L_{fpo} must be at a distance past unstable ground sections, with the extent of tension failure assumed. The axisymmetric analysis was found to provide a quick approximation of this design parameter.

The coverage angle of the forepole elements, α_{fpa}, was found to exert a greater influence on the global response of a system when compared with the centre to centre spacing of the forepole elements, S_{cfp}. The S_{cfp} was found to be a critical design parameter, yet one that could not be easily captured in full scale 3D numerical analysis. The authors have therefore proposed a 2D analysis approach in order to capture the maximum spacing, based on typical size/stiffness of the forepole elements (diameter, ϕ_{fp}, and thickness, t_{fp}), and the angle of installation α_{fp}. Such approaches can be employed for both squeezing ground conditions (continuum model) and gravity driven failure (particle, discrete, and jointed continuum models).

The increase of the angle of installation from the horizontal was found to slightly decrease the convergence of the continuous profile of the tunnel excavation. This result, however, proved to be relatively insignificant for a "saw-tooth" profile excavation due to the increased requirements of other support members. As noted previously, the umbrella arch with forepole elements is always employed in conjunction with other temporary support systems; these additional systems also have an influence on the global response of the complete system that must also be determined.

In conclusion, this paper presented the use of numerical techniques in order to add value to the understanding of the influences of design parameters on forepole elements used within an umbrella arch system. The paper also provided an overview of suggested, sound numerical modelling procedures for support systems, optimisation of design, and the comprehensive performance of full 3D analysis of preliminary design.

It should be noted, however, that sound engineering judgement and a comprehensive understanding of the fundamentals associated with geotechnical, mechanical and numerical analysis factors are always prerequisites prior to conducting a preliminary design for tunnel construction of this nature. A designer must understand the limitations associated with the numerical tools combined with the accuracies related to obtaining sound geotechnical data (i.e. input and interaction parameters, site-specific factors, etc.).

ACKNOWLEDGEMENT

This work has been funded by the Natural Sciences and Engineering Research Council of Canada, the Department of National Defence (Canada) as well as graduate funding obtained at Queen's University and the Royal Military College of Canada. A special thanks goes to Ehsan Ghazvinian for his help with the Voronoi and PFC considerations.

REFERENCES

1. Broch E, Lu M, Trinh QN. Three dimensional modelling of a tunnel cave-in and spiling bolt support. In: Leung CF, Zhou YX, editors. Proceedings of the ISRM international symposium 2006 and the 4th Asian rock mechanics symposium. Singapore: World Scientific Publishing Co., Pte. Ltd; 2006.
2. Carranze-Torres C. Elasto-plastic solution of tunnel problems using the generalized form of the Hoek-Brown failure criterion. International Journal of Rock Mechanics and Mining Sciences 2004;41(3):480e91.
3. Carrieri G, Grasso P, Mahtab A, Pelizza S. Ten years of experience in the use of umbrella-arch for tunneling. In: Proceedings of the international congress on soil and rock improvement in underground works; 1991. p. 99e111.
4. Doi Y, Otani T, Shinji M. The optimum distance of roof umbrella method for soft ground by using PFC. In: Ma GW, Zhou YX, editors. Analysis of Discontinuous Deformation: New Developments and Applications. Singapore: Research Publishing Services; 2009. p. 461e8.
5. Egnatia Odos SA. Geological study by asproudas and cooperates consultants. 1998 (in Greek). Egnatia Odos SA. The appropriate use of geological information in the design and construction of the Egnatia motorway tunnels, tunnel construction in Greece. Technical notes of the National Technical University of Athens. 2001.
6. Federal Highway Administration (FHA). Technical manual for design and construction of road tunnelsecivil elements. 2009. http://www.fhwa.dot.gov/Bridge/tunnel/pubs/nhi09010/errata.cfm.
7. Funatsu T, Hoshino T, Sawae H, Shimizu N. Numerical analysis to better understand the mechanism of the effects of ground supports and reinforcements on the stability of tunnels using the distinct element method. Tunnelling and Underground Space Technology 2008;23(5):561e73.

8. Grasso P, Scotti G, Blasini G, Pescara M, Floria V, Kazilis N. Successful application of the observational design method to difficult tunnel conditionseDriskos tunnel. In: Erdem Y, Solak T, editors. Underground Space Use: Analysis of the Past and Lessons for the Future. 2005. p. 463e70.
9. Hoek E. Big tunnels in bad rock: 2000 Terzaghi lecture. ASCE Journal of Gotechnical and Geoenvironmental Engineering 2001;127(9):726e40.
10. Hoek E. Rock-support interaction analysis for tunnels in weak rock masses. In: Evert H, editor. Practical Rock Engineering. Rocscience; 2007.
11. Hoek E. Support for very weak rock associated with faults and shear zones. In: International symposium of rock and support and reinforcement practice in mining. Kalgoorlie, Australia; 1999.
12. Hun Y. Stability and collapse mechanism of unreinforced and forepole reinforced tunnel headings. PhD Thesis. Singapore: National University of Singapore; 2011. Itasca Consulting Group Inc. Fast Lagrangian analysis of continua in 3 dimensions, version 4.0. Minneapolis: Itasca Consulting Group Inc.; 2009.
13. Itasca Consulting Group Inc. Particle flow code, PFC v3.0. Minneapolis: Itasca Consulting Group Inc.; 2002.
14. Kim SH, Baek SH, Moon HK. A study on the reinforcement effect of umbrella arch method and prediction of tunnel crown and surface settlement. In: Erdem Y, Solak T, editors. Underground Space Use: Analysis of the Past and Lessons for the Future. London: Taylor & Francis; 2005. p. 245e51.
15. Mager W, Mocivnik J. Modern casing technology sets a milestone in drilling and ground anchoring. Felsbau 2000;18:43.
16. Marinos V, Fortsakis P, Prountzopoulos G. Estimation of rock mass properties of heavily sheared flysch using. In: IAEG2006. The Geological Society of London; 2006. Paper No.314. Ocak I. Control of surface settlement with umbrella arch method in second stage excavation of Istanbul metro. Tunnelling and Underground Space Technology 2008;23(6):674e81.
17. Oke J, Vlachopoulos N, Diederichs MS. Improved input parameters and numerical analysis techniques for temporary support of underground excavations in weak rock. In: RockEng. Edmonton; 2012.
18. Oke J, Vlachopoulos N, Diederichs MS. Modification of the supported longitudinal displacement profile for tunnel face convergence in weak rock. In: 47th US rock mechanics/geomechanics symposium. San Francisco: American Rock Mechanics Association; 2013b.
19. Oke J, Vlachopoulos N, Diederichs MS. Semi-analytical model of an umbrella arch employed in hydrostatic tunnelling conditions. In: 48th US rock mechanics/ geomechanics symposium. Minneapolis: American Rock Mechanics Association; 2014b.

20. Oke J, Vlachopoulos N, Diederichs MS. The reduction of surface settlement by employing umbrella arch systems. In: GeoMontreal 2013. Montreal: Canadian Geotechnical Society; 2013a.
21. Oke J, Vlachopoulos N, Diederichs MS. The reduction of surface settlement by employing umbrella arch systems for different excavation methods. In: The ISRM European rock mechanics symposiumeEUROCK2014. Vigo: International Society of Rock Mechanics; 2014c.
22. Oke J, Vlachopoulos N, Marinos V. The pre-support nomenclature and support selection methodology for temporary support systems within weak rock masses. Journal of Geotechnical and Geological Engineering 2014a;32(1):97e130.
23. Otani T, Shin JI, Chijiwa T. Proposal of numerical model and the determination method of design parameters for pipe roofing method. Daboku Gakkai Ronbunshuu 2008;64(4):450e62.
24. Peila D. Forepoling design. In: Ground Improvement, Pre-support and Reinforcement Short Course. Geneva: International Tunnelling and Underground Space Association (WTC 2013); 2013.
25. Potyondy DO, Cundall PA. A bonded-particle model for rock. International Journal of Rock Mechanics and Mining Sciences 2004;41(8):1329e64.
26. Rocscience Inc.. Phase2 v7. Toronto, Canada: Rocscience Inc; 2010.
27. Rocscience Inc.. Phase2 v8. Toronto, Canada: Rocscience Inc.; 2013.
28. Shin JH, Choi YK, Kwon OY, Lee SD. Model testing for pipe-reinforced tunnel heading in a granular soil. Tunnelling and Underground Space Technology 2008;23(3):241e50.
29. Song KI, Cho GC, Chang SB, Lee IM. Beam-spring structural analysis for the design of a tunnel pre-reinforcement support system. International Journal of Rock Mechanics and Mining Sciences 2013;59:139e50.
30. St. John CM, Van Dillen DE. Rockbolts: a new numerical representation and its application in tunnel design. In: 24th US symposium on rock mechanics. New York: Association of Engineering Geologists; 1983. p. 13e26.
31. Stockl C. Numerische Berechnung der Tragwirkung von Rohrschirmen mit PFC-2D. MS Thesis. Graz: Graz University of Technology; 2002 (in Germany).
32. Trinh QN. Analyses of a cave-in problem in a hydropower tunnel in Vietnam. PhD Thesis. Trondheim, Norway: Norwegian University of Science and Technology; 2006.
33. Vlachopoulos N, Diederichs MS, Marinos V, Marinos P. Tunnel behaviour associated with the weak Alpine rock masses of the Driskos twin tunnel system, Egnatia Odos Highway. Canadian Geotechnical Journal 2013;50(1):91e120.

34. Vlachopoulos N, Diederichs MS. Appropriate uses and practical limitations of 2D numerical analysis of tunnels and tunnel support response. Geotechnical and Geological Engineering 2014;32(2):469e88.
35. Vlachopoulos N. Back analysis of a tunnelling case study in weak rock of the alpine system in northern Greece: validation and optimization of design analysis based on ground characterization and numerical simulation. PhD Thesis. Kingston, Ontario: Queens University; 2009.
36. Volkmann G. Rock mass-pipe roof support interaction measured by chain inclinometers at the Birgltunnel. In: Proceedings of international symposium on geotechnical measurements and modeling. Karlsruhe: A.A. Balkema; 2003. p. 105e9.
37. Volkmann GM, Dolsak W. Development of innovative connections for pipe umbrella support systems. In: Tunnels for a Better Life, Proceedings of the World Tunnel Congress. Iguassu Falls, Brazil; 2014.
38. Volkmann GM, Schubert W, Button EA. A contribution to the design of tunnels supported by a pipe roof. In: 50 Years of rock mechanicsdLandmarks and future challenges, Proceedings of the 41st US symposium on rock mechanics. Golden, Colorado; 2006.
39. Volkmann GM, Schubert W. A load and load transfer model for pipe umbrella support. In: Zhao J, Labiouse V, Dudt JP, Mathier JF, editors. Rock mechanics in civil and environmental engineering, proceedings of the European rock mechanics symposium (EUROCK) 2010. Rotterdam, Niederlande: CRC Press/A.A. Balkema; 2010. p. 379e82.
40. Volkmann GM, Schubert W. Contribution to the design of tunnels with pipe roof support. In: Rock mechanics in underground construction: ISRM international symposium 2006: 4th Asian rock mechanics symposium. Singapore: World Scientific Pub. Co. Inc.; 2006a.
41. Volkmann GM, Schubert W. Geotechnical model for pipe roof supports in tunneling. In: Proceeding of the 33rd ITA-AITES World tunneling congress, underground spacedthe 4th dimension of metropolises. London: Taylor & Francis Group; 2007. p. 755e60.
42. Volkmann GM, Schubert W. Optimization of excavation and support in pipe roof supported tunnel sections. Tunnelling and Underground Space Technology 2006b;21(3e4):404.
43. Wang H, Jia J. Analytical method for mechanical behaviors of pipe roof reinforcement. In: 2008 International conference on information management, innovation management and industrial engineering. Institute of Electrical and Electronics Engineers (IEEE); 2008. p. 352e7.

44. Yasitli NE. Numerical modeling of surface settlements at the transition zone excavated by New Austrian tunneling method and umbrella arch method in weak rock. Arabian Journal of Geosciences 2013;6(7):2699e708.

CITATION

J. Oke, N. Vlachopoulos, M.S. Diederichs, Numerical analyses in the design of umbrella arch systems, Journal of Rock Mechanics and Geotechnical Engineering, Volume 6, Issue 6, December 2014, Pages 546-564, ISSN 1674-7755, http://dx.doi.org/10.1016/j.jrmge.2014.09.005.

CHAPTER 8

Fracture Initiation and Propagation in Intact Rock – A Review

E. Hoeka, C.D. Martinb

[a]West Vancouver, British Columbia, V7V 0B3, Canada

[b] Department of Civil & Environmental Engineering, University of Alberta, Edmonton, Alberta, T6G 2W2, Canada

ABSTRACT

The initiation and propagation of failure in intact rock are a matter of fundamental importance in rock engineering. At low confining pressures, tensile fracturing initiates in samples at 40%–60% of the uniaxial compressive strength and as loading continues, and these tensile fractures increase in density, ultimately coalescing and leading to strain localization and macro-scale shear failure of the samples. The Griffith theory of brittle failure provides a simplified model and a useful basis for discussion of this process. The Hoek–Brown failure criterion provides an acceptable estimate of the peak strength for shear failure but a cutoff has been added for tensile conditions. However, neither of these criteria adequately explains the progressive coalition of tensile cracks and the final shearing of the specimens at higher confining stresses. Grain-based numerical models, in which the grain size distributions as well as the physical properties of the component grains of the rock are incorporated, have proved to be very useful in studying these more complex fracture processes.

INTRODUCTION

In order to understand the characteristics of rock and rock masses as engineering materials, it is necessary to start with the behavior of intact rock. From an engineering point of view, this involves studying laboratory-scale samples, such as diamond drill core, with dimensions in the range of 50 mm diameter. For many rock types, the grain size is small enough that samples of this scale can be considered homogeneous and isotropic.

The characteristics that will be discussed in the following text are the strength and deformation characteristics of intact rock. As illustrated in Fig. 1, a number of stress states need to be considered and, as is common in most discussions on this topic, it will be assumed that these stress states can be considered in two dimensions. In other words, it is assumed that the intermediate principal stress σ_2 has a minimal influence on the initiation and propagation of failure in the samples. While some authors consider this to be an over-simplification, a full three-dimensional treatment of the topic would result in complex text which would defeat the purpose of this presentation which is designed to be as clear and understandable as possible.

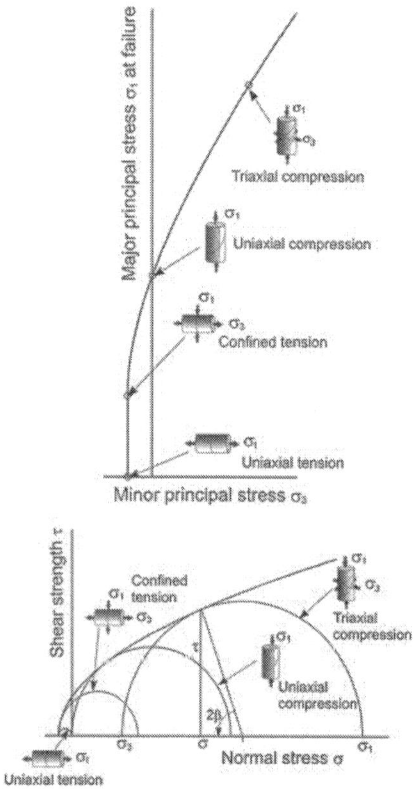

Figure 1. Typical failure characteristics of intact rock plotted in terms of major and minor principal stresses and Mohr circles and envelope.

THEORETICAL FRACTURE INITIATION: BACKGROUND

Griffith tensile theory

Griffith (1921) proposed that tensile failure in brittle materials such as glass initiates at the tips of minute defects which he represented by flat elliptical cracks. His original work dealt with fracture in material subjected to tensile stress but later he extended this concept to include biaxial compression loading (Griffith, 1924). The equation governing tensile failure initiation in a biaxial compressive stress field is

$$\sigma_1 = \frac{-8\sigma_t\left(1+\frac{\sigma_3}{\sigma_1}\right)}{\left(1-\sigma_3/\sigma_1\right)^2} \qquad (1)$$

where σ_t is the uniaxial tensile strength of the material. Note that tensile stresses are negative.

Murrell (1958) proposed the application of Griffith theory to rock. In the 1960s, Griffith's two-dimensional theory was extended to three dimensions by a number of authors including Murrell (1958), Sack and Kouznetsov whose work was summarized in books on brittle failure of rock materials by Andriev (1995) and Paterson and Wong (2005). These extensions involve examining the stresses induced around open penny-shaped cracks in a semi-infinite body subjected to triaxial compressive stresses σ_1, σ_2 and σ_3. It was shown that the intermediate principal stress σ_2 has no significant influence on the crack tip stresses inducing tensile failure initiation. Hence, this criterion is essentially equivalent to loading a penny-shaped crack in a biaxial stress field, as shown in Fig. 2.

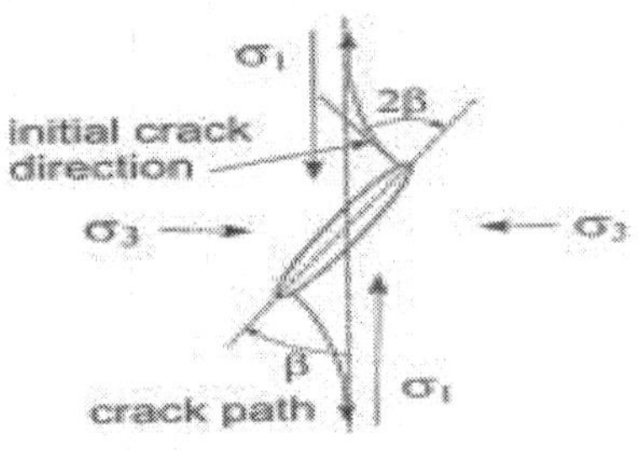

Figure 2. Tensile crack propagation from an inclined elliptical Griffith crack in a biaxial compressive stress field.

The equation governing tensile failure initiation is

$$\sigma_1 = \frac{-12\sigma_t\left(1 + 2\dfrac{\sigma_3}{\sigma_1}\right)}{\left(1 - \sigma_3/\sigma_1\right)^2} \tag{2}$$

Note that, whereas the original Griffith theory predicts a ratio of compressive to tensile strength σc/|σt|=8, the penny-shaped crack version predicts σc/|σt|=12. The corresponding Mohr envelope for the penny-shaped crack version is

$$\tau^2 = |\sigma_t|(|\sigma_t| + \sigma)\left(\sqrt{\frac{\sigma_c}{|\sigma_t|} + 1} - 1\right)^2 \tag{3}$$

where σ_c is the uniaxial compressive strength of the material.

The Griffith theory deals only with the initiation of tensile failure. It cannot be extended to deal with failure propagation and eventual shear failure in compression. However, under certain conditions when tensile stresses exceed the tensile strength, tensile failure initiation can lead to crack propagation. In these cases the tensile cracks propagate along the major principal stress (σ_1) trajectory as shown in Fig. 2.

Modifications to Griffith theory for closed cracks

The original Griffith theory was derived from analyses of crack initiation at or near the tips of open elliptical cracks. In the case of rocks, most of the defects from which tensile cracks originate are grain boundaries which are usually cemented and have to be considered as closed cracks. McClintock and Walsh (1962) proposed that tensile fracture from closed Griffith cracks can be predicted on the basis of the conventional Mohr–Coulomb equation: where ϕ is the angle of friction and τ_0 is the shear strength at zero normal stress.

$$T = \tau_0 + \sigma \tan\phi \tag{4}$$

Hoek (1965) discussed the transition from the Griffith theory for open cracks, which applies for confining stresses $\sigma_3 < 0$, and the modified theory for closed cracks which applies for compressive confining stresses. For the principal stress plot, this transition occurs at $\sigma_3 = 0$, while for the Mohr envelope, the transition occurs at the tangent points on the Mohr circle representing the uniaxial compressive strength σ_c of the intact rock. The transition is illustrated in Fig. 3 in which the principal stress plots are shown for friction angles of 35°, 45° and 55°.

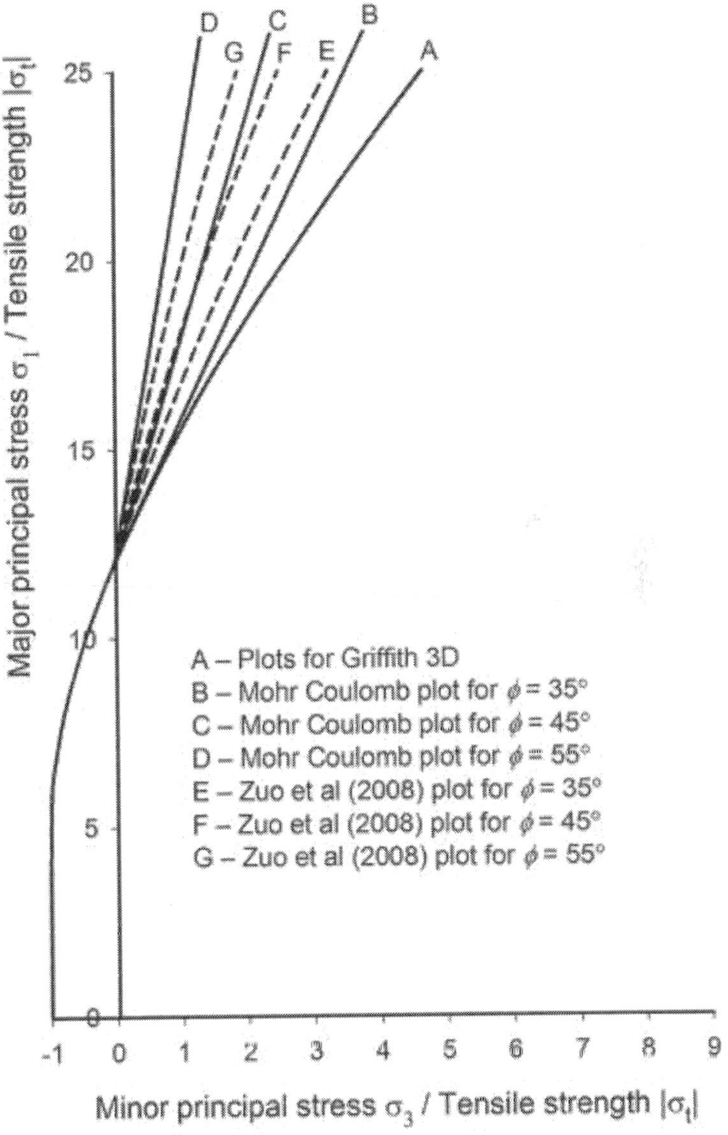

Figure 3 Principal stress plots for various criteria for tensile failure initiation from closed cracks in brittle materials such as rock.

A much more comprehensive discussion on this topic is given in Paterson and Wong (2005) but the plotted results are essentially the same as those shown in Fig. 3. Hence, for the purpose of this discussion, Eq. (4) above is adequate.

Zuo et al. (2008) examined the growth of microcracks in rock-like materials on the basis of fracture mechanics considerations. They assumed a sliding-crack model which generates wing cracks, similar to those shown in Fig. 2, from close to the crack tips when the frictional strength of the sliding surfaces is overcome. They found that the failure initiation criterion can be expressed by the following equation:

$$\sigma_1 = \sigma_3 + \sqrt{\frac{\mu}{\kappa}\frac{\sigma_c}{|\sigma_t|}\sigma_c\sigma_3 + \sigma_c^2} \tag{5}$$

where μ is the coefficient of friction which is equal to the tangent of the friction angle, i.e. $\mu = \tan\phi$.

The coefficient κ is used for mixed mode fracture and it can be derived from various approximations based on a maximum stress criterion or a maximum energy release criterion (Zuo et al., 2008). Plots for Eq. (5), when $\mu = 0.7$, 1 and 1.43 ($\phi = 35°$, 45° and 55°), $\kappa = 1$ and $\sigma c/|\sigma t| = 12$, are included in Fig. 3. Note that the same transition from open to closed crack behavior has been assumed as for the Mohr–Coulomb criterion (Eq. (4)) discussed above.

Length of induced tensile cracks

Hoek (1965) carried out experiments in which flat open "cracks" were machined ultrasonically into annealed glass plates which were then loaded biaxially. The initiation of tensile cracks from near the tips of these simulated cracks, as predicted by Griffith's original theory, was confirmed. However, it was found that the length of the tensile cracks was limited by the ratio of the applied biaxial stresses σ_3/σ_1. As reported by Cho et al. (2007), theoretical studies on closed cracks have been carried out by several authors including Ashby and Hallam, 1986, Kemeny and Cook, 1987, Germanovich and Dyskin, 1988 and Martin, 1997 and Cai et al. (1998). These studies, the results of which are plotted in Fig. 4, confirm the importance of confinement in limiting the length of induced tensile cracks from pre-existing flaws in brittle materials subjected to compressive loading. Fig. 5 summarizes some of this information in a different form and shows a principal stress plot and Mohr's diagram for open penny-shaped cracks subjected to different biaxial compressive stress loadings.

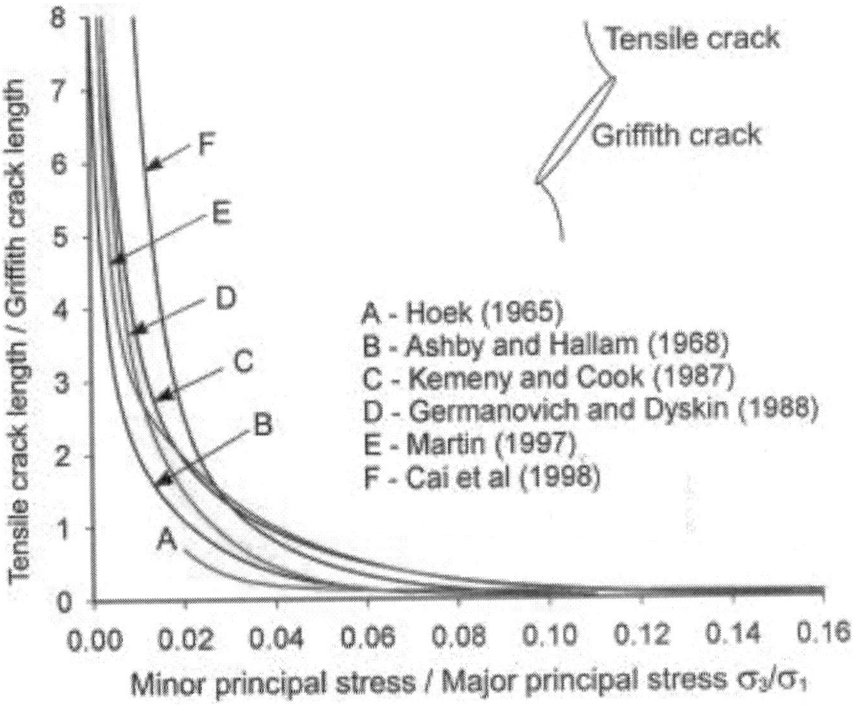

Figure 4. Dependence of length of tensile cracks on principal stress ratio σ_3/σ_1.

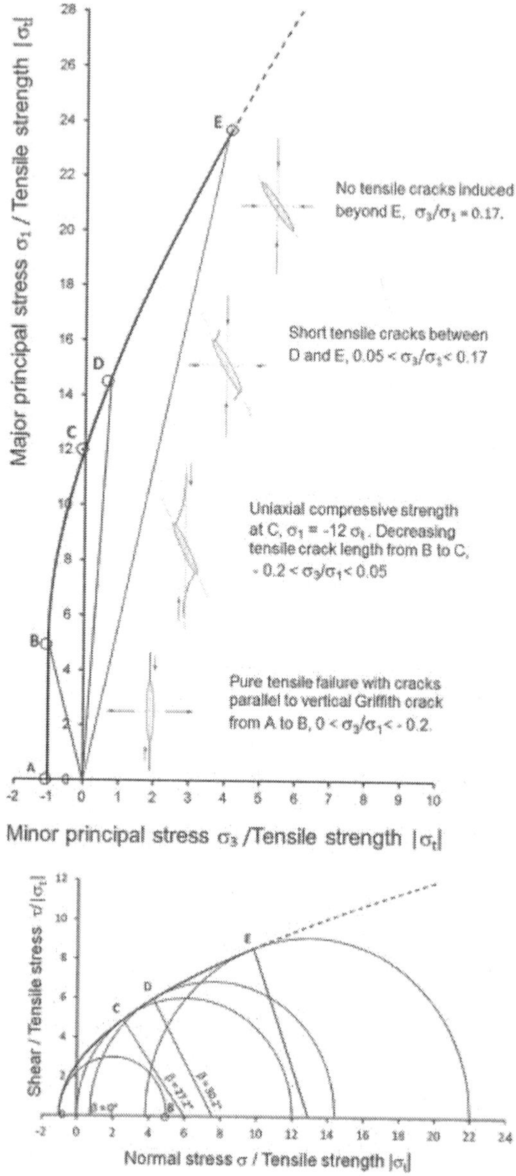

Figure 5. Plots of principal stresses defining tensile failure initiation from open penny-shaped cracks in a homogeneous isotropic elastic solid loaded biaxially.

Summary

Griffith theory of brittle fracture initiation and its modifications have been discussed in hundreds of technical papers. A particularly useful review was presented by Fairhurst (1972) which is recommended reading for anyone interested in pursuing this topic in greater depth. While there can be no dispute that this is very important background material for an understanding of the mechanics of brittle failure initiation, it is of limited practical value in the field of rock engineering.

This is because an isolated Griffith crack in a semi-infinite plate is an inadequate model of the grain boundary network in which tensile failure originates and propagates in intact rock as shown in Fig. 6. This photograph shows that, while it would always be possible to find a critically oriented grain boundary from which tensile failure could initiate, it is unlikely that the induced crack would follow the path suggested in Fig. 2 for homogeneous isotropic materials. Rather, the tensile crack path would follow a path dictated by grain boundaries with only isolated cracks running across intact grains. This means that we have to explore other more complicated models in order to fully understand the fracture process in rock.

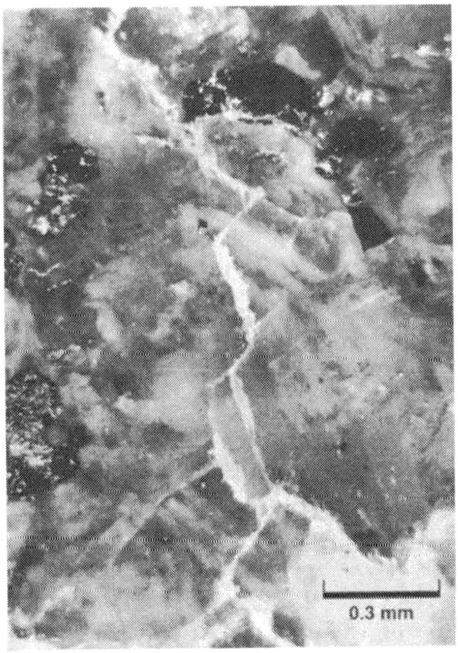

Figure 6. Crack path in a specimen of Witwatersrand Quartzite from a deep gold mine in South Africa, sectioned after uniaxial compressive loading to about 75% of the uniaxial compressive strength. Photograph reproduced from Hoek and Bieniawski (1965).

Before leaving the topic of Griffith theory and its modifications, it is worth summarizing what we have learned from the discussion given earlier since the same or similar issues will apply to the numerical analysis of fracture initiation and propagation:

1) The brittle failure process initiates and is, to a very large extent, controlled by the tensile strength of intact rock or of its component grains.
2) The initiation of tensile cracks at or near the tip of a Griffith crack, whether this crack is open or closed, depends upon the orientation of the Griffith crack in relation to the applied stresses. Fracture will initiate at or close to the tip of a critically oriented crack when the conditions defined by Eqs. (1), (2), (3) and (5) are satisfied, depending upon the assumptions made in deriving these equations.
3) This process is extremely sensitive to the degree of confinement and the extent of failure reduces quickly as the minor principal stress (σ_3) increases from $\sigma_3 = \sigma_t$ to$\sigma_3 > 0$, as shown in Fig. 1 and Fig. 5.
4) At some level of confinement, in the range of $\sigma_3/\sigma_1 \approx 0.2$, tensile failure can be suppressed completely and the peak strength of the intact rock is controlled by shear failure for higher confinement.
5) For applications to confined rock materials, the closed Griffith crack model (Eqs. (4) and (5)) is the most appropriate. The shear strength of the confined defects (typically grain boundaries) is a controlling parameter in the initiation and propagation of the tensile failure.
6) The ratio of uniaxial compressive to tensile strength ($\sigma c/|\sigma t|$) is an important parameter in understanding the failure of rock and similar brittle materials.

FRACTURE INITIATION AND PROPAGATION: LABORATORY TESTS

Peak strength and the Hoek–Brown criterion

Hoek and Brown (1980) and Hoek (1983) described the development of the Hoek–Brown failure criterion as a trial-and-error process using the Griffith theory as a starting point. They were seeking an empirical relationship that fitted observed shear failure conditions for brittle rock subjected to triaxial compressive stresses. The equation chosen to represent the failure of intact rock was

$$\sigma_1 = \sigma_3 + \sigma_c\sqrt{m_i\frac{\sigma_3}{\sigma_c} + 1} \tag{6}$$

where *mi* is a material constant.

Zuo et al. (2008) pointed out that the substitution of (μ/κ) (σc/|σt|) = mi in their failure criterion (Eq. (5)) leads to the Hoek–Brown criterion for intact rock (Eq. (6)). They suggested that the constant *mi* is not simply an empirical constant but that it has real physical meaning.

During the 1970s, when the Hoek–Brown criterion was developed, there was little interest in the tensile strength of rock and, in fact, it was frequently assumed to be zero. The emphasis was on confined shear failure which was assumed to control the stability of the relatively small slopes and shallow tunnels that were constructed at the time. However, with the increase in depth of excavations in civil and mining engineering projects and the depth of boreholes in oil exploration and recovery, the issue of the tensile strength of rock became increasingly important. In particular, the process of brittle fracture which results in splitting, popping, spalling and rockbursting in pillars and tunnels, and "breakouts" in boreholes is a tensile failure process which is not adequately dealt with by the Hoek–Brown failure criterion. Simply projecting the Hoek–Brown equation (Eq. (6)) back to its σ_3 intercept with $\sigma_1 = 0$ does not give an acceptable value for the tensile strength of the rock.

Ramsey and Chester (2004) and Bobich (2005) have investigated this issue in a series of experiments in which they used dogbone-shaped specimens as shown in Fig. 7. By choosing appropriate diameters for the ends and center of the specimen and by adjusting the values of the confining pressure P_c and the axial stress P_a, a range of values of $\sigma_3$and σ_1 can be generated in the test section.

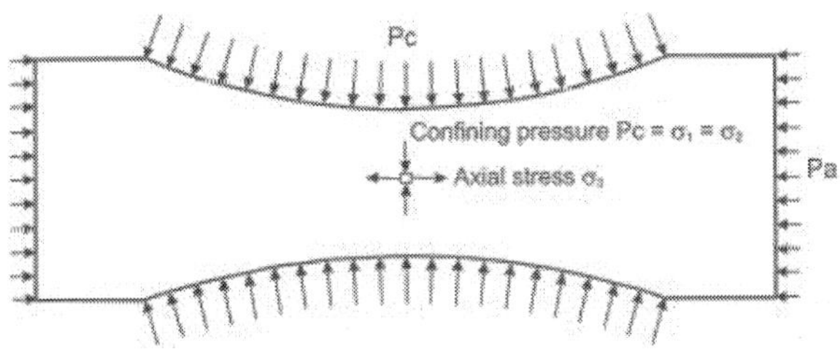

Figure 7. Dogbone-shaped specimen used by Ramsey and Chester (2004) to investigate tensile failure of Carrara marble.

The results of tests on Carrara marble are reproduced in Fig. 8. The Hoek–Brown criterion (Eq. (6)) has been fitted to the shear data obtained in these tests and the resulting curve has been projected back to give an intercept of $\sigma_3 = -17.2$ MPa for $\sigma_1 = 0$. As can be seen in Fig. 8, this does not correspond to the tensile failure data which gives an average tensile strength $\sigma_3 = -7.75$ MPa. In other words, the Hoek–Brown criterion has no provision for predicting the tensile strength shown in Fig. 8, and highlighted by the "Tension cutoff".

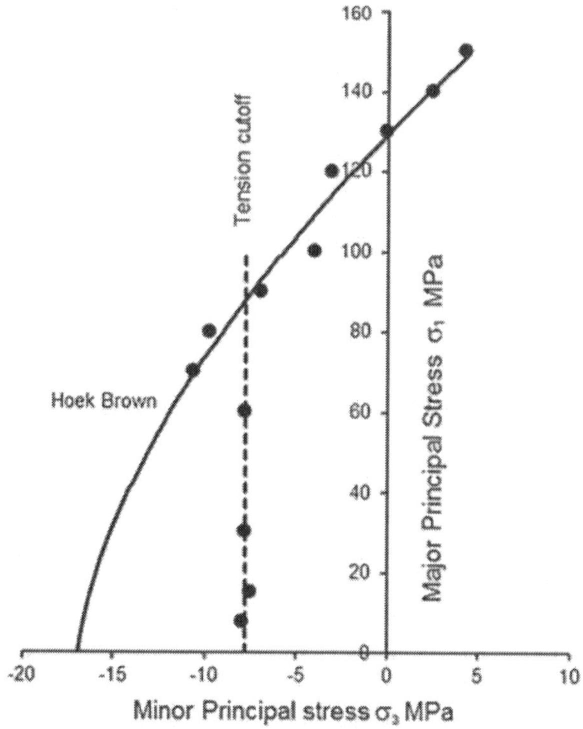

Figure 8. Results from confined extension tests and triaxial compression tests by Ramsey and Chester (2004).

Fairhurst's generalized Griffith fracture criterion

Fairhurst (1964) proposed that the Griffith failure criterion, discussed in Section 2 of this paper, could be generalized in terms of the ratio of compressive to tensile strength $\sigma c/|\sigma t|$ as follows (a detailed derivation is given in the Appendix):

1) If $w(w-2)\,\sigma_3 + \sigma_1 \leq 0$, failure occurs when $\sigma_3 = \sigma_t$;

2) If $w(w-2)\,\sigma_3 + \sigma_1 \leq 0$, failure occurs when

$$\sigma_1 = \frac{(2\sigma_3 - A\sigma_t) + \sqrt{(A\sigma_t - 2\sigma_3)^2 - 4\left(\sigma_3^2 + A\sigma_t\sigma_3 + 2AB\sigma_t^2\right)}}{2} \tag{7}$$

$$\left.\begin{aligned} A &= 2(w-1)^2 \\ B &= [(w-1)/2]^2 - 1 \\ w &= \sqrt{\sigma_c/|\sigma_t| + 1} \end{aligned}\right\} \tag{8}$$

Fitting Eq. (7) to the results plotted in Fig. 8 gives the combined plot shown in Fig. 9.

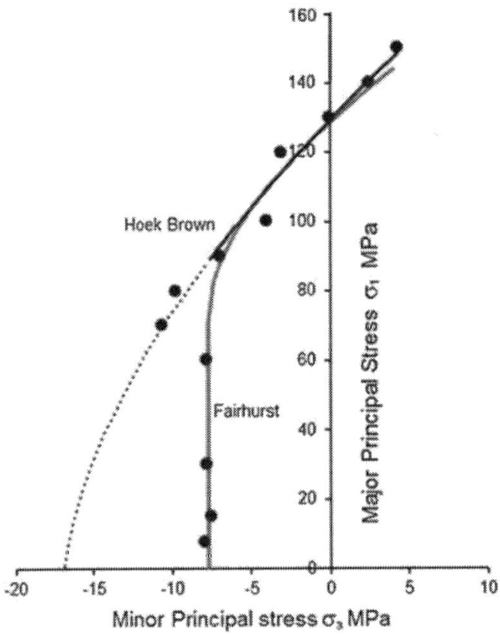

Figure 9. Combined plot of Hoek–Brown and Fairhurst failure criteria with tension cutoff.

Reliable direct tensile test data on rock are very rare and the authors have only been able to assemble the limited number of results included in Table 1. However, by fitting both Hoek–Brown and Fairhurst curves to these data, as shown in Fig. 9, it has been possible to arrive at a preliminary relationship between the Fairhurst tension cutoff (defined by$\sigma c/|\sigma t|$) and the Hoek–Brown parameter *mi* plotted in Fig. 10. While more work remains to be done on this topic, particularly more tests of the type carried out by Ramsey and Chester (2004) and Bobich (2005), the authors suggest that Fig. 10 provides a useful practical tool for estimating a tensile cutoff for the Hoek–Brown criterion.

Table 1. Analysis of data containing reliable tensile values.

Fairhurst ($\sigma_t \leq 2$ MPa)			Hoek–Brown (shear data)			Data set and reference
σ_c(MPa)	σ_t(MPa)	$\sigma c/\|\sigma t\|$	σ_c(MPa)	σ_t(MPa)	mi	
128.5	-7.74	16.6	129	−15.6	8.25	Carrara marble (Ramsey and Chester, 2004)
516.5	−33.72	13.9	557	−65.9	8.45	Blair dolomite (Brace, 1964)
95.5	−6.41	14.9	102	−10.6	9.65	Berea sandstone (Bobich, 2005)
125.5	−8.72	14.4	131	−12.4	10.6	Webtuck dolomite (Brace, 1964)
614	−25.5	24.1	592	−28.7	20.65	Granite aplite (Hoek, 1965)
220	−7.06	31.1	227	−6.01	32.4	Lac du Bonnet granite (Lau and Gorski, 1992)

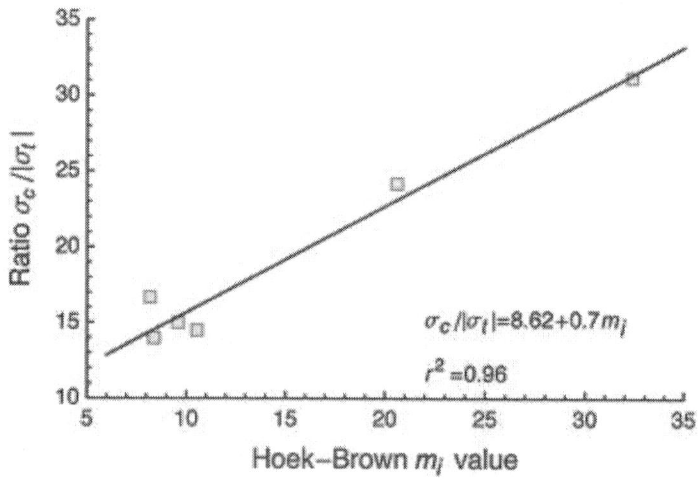

Figure 10. Relationship between $\sigma c/|\sigma t|$ and *mi* from Table 1.

Examination of Table 1 shows that, for low *mi* values, the Hoek–Brown criterion over-estimates the tensile strength compared with the Fairhurst criterion. However, for *mi* > 25 the Hoek–Brown criterion under-estimates the tensile strength by an amount that is generally small enough to be ignored for most engineering applications.

Hoek (1965) assembled a significant quantity of laboratory triaxial test data for a variety of rock types and concrete and these results (peak strength values) are plotted in a dimensionless form in Fig. 11. It can be seen that individual data sets plot on parabolic curves and that a family of such curves, covering all of the shear data collected, can be generated for different values of the Hoek–Brown constant *mi*. The constant *mi* is an indicator of the brittleness of the rock with weaker and more ductile rocks having low *mi*values while stronger and more brittle rocks have high *mi* values. A few data points for$\sigma_3 < 0$ are included in Fig. 11 and these are dealt with adequately by the tension cutoff discussed above.

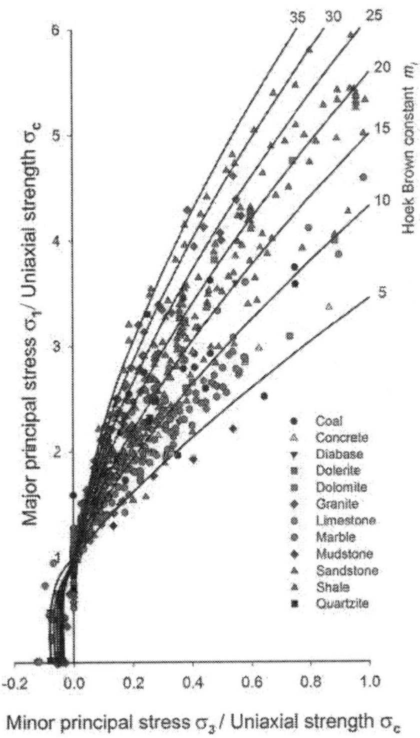

Figure 11. Dimensionless plot of triaxial test results from laboratory tests on samples from a wide range of rock types and concrete.

Interpretation of laboratory triaxial tests

In order to understand the application of the Hoek–Brown failure criterion to intact rock behavior, it is useful to consider a practical example which involves uniaxial and triaxial tests on specimens of Lac du Bonnet granite from the site of the Atomic Energy of Canada Limited Underground

Research Laboratory at Pinawa in Manitoba, Canada (Read and Martin, 1991). The specimen preparation and testing were carried out by the CANMET Mining and Mineral Sciences Laboratories in Ottawa, Canada, which has a long-standing international reputation for high quality testing services.

Using strain and acoustic emission measurements, Lau and Gorski (1992) determined the crack initiation, onset of strain localization and peak strengths for each confining pressure in a series of triaxial tests carried out in a servo-controlled stiff testing machine, based on the procedure summarized in Fig. 12 (Martin and Chandler, 1994). The results of these tests are plotted in Fig. 13. The Hoek–Brown failure criterion (Eq. (6)) has been fitted to each data set and the fitted parameters are included in Table 2. Note that, because the *mi* value for the peak strength is 32.4, no correction has been made for the measured tensile strength, as discussed above.

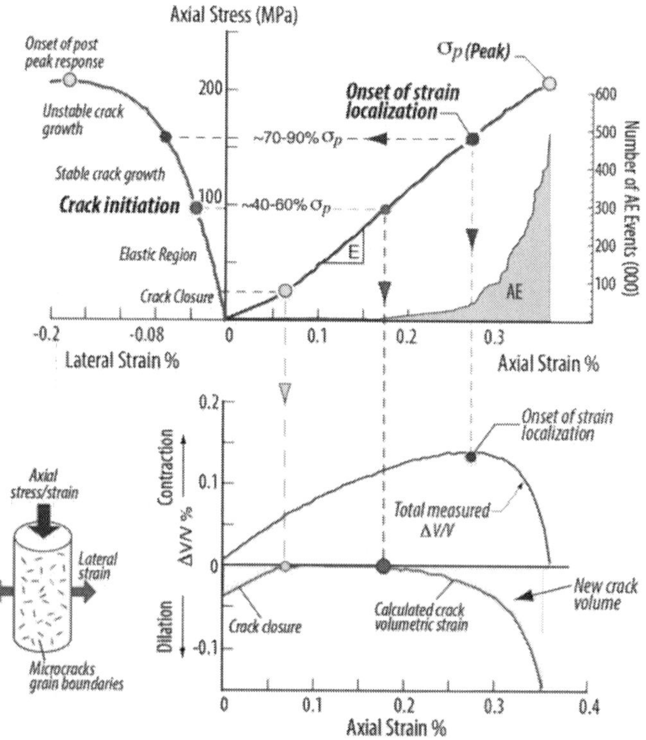

Figure 12. Stages in the progressive failure of intact rock specimens subjected to compressive loading. Modified from Martin and Christiansson (2009).

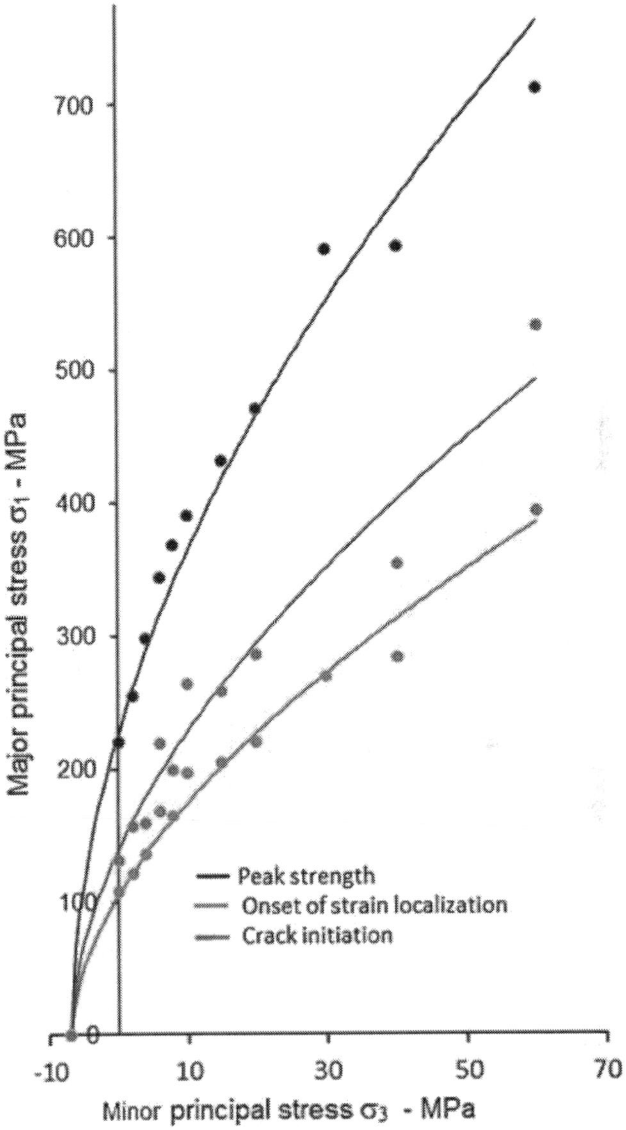

Figure 13. Tensile crack initiation, strain localization and peak strength for Lac du Bonnet granite from tests by Lau and Gorski (1992).

Table 2. Results of triaxial tests on Lac du Bonnet granite.

Confining stress (MPa)	Crack initiation (MPa)	Strain localization (MPa)	Peak strength (MPa)	Average tensile strength (direct tension) (MPa)
0	131	108	220	−7
2	157	121	255	
4	159	136	298	
6	219	168	344	
8	199	165	368	
10	264	197	391	
15	258	205	432	
20	286	220	471	
30	–	269	591	
40	354	284	593	
60	533	394	712	
Hoek–Brown parameters used	σ_c = 106 MPa,*mi* = 15	σ_c = 140 MPa,*mi* = 20	σ_c = 227 MPa,*mi* = 32.4	

It is clear from Fig. 12 and Fig. 13 that fracturing in laboratory samples is a complex process and that simply measuring the peak stress does not capture this fracturing process. However, it is also clear from Fig. 13 that we can define the boundaries for this process, i.e. fracture initiation, onset of fracture localization and collapse peak stress. In the next section, a numerical approach that can simulate this process is examined.

Numerical approaches

Since the early 1980s, there has been an exponential growth in the sophistication and power of numerical programs which have been increasingly applied to the study of failure initiation and propagation processes in soil, rock and concrete. Fig. 14 illustrates two phenomenological approaches that are typically used to replicate the failure process numerically. The early approaches often used the sliding-crack model to capture many of the elements discussed in the earlier section on Griffith theory. More recently, there has been an increasing focus on the force-chain crack model using discrete element formulations

(Fig. 14). A small selection of some of the more significant papers in this latter field includes: Cundall and Strack, 1979, Diederichs, 1999, Diederichs, 2003,Potyondy and Cundall, 2004, Pierce et al., 2007, Lorig, 2007, Cho et al., 2007, Cundall et al., 2008, Lan et al., 2010 and Potyondy, 2012 and Scholtès and Donzé (2013).

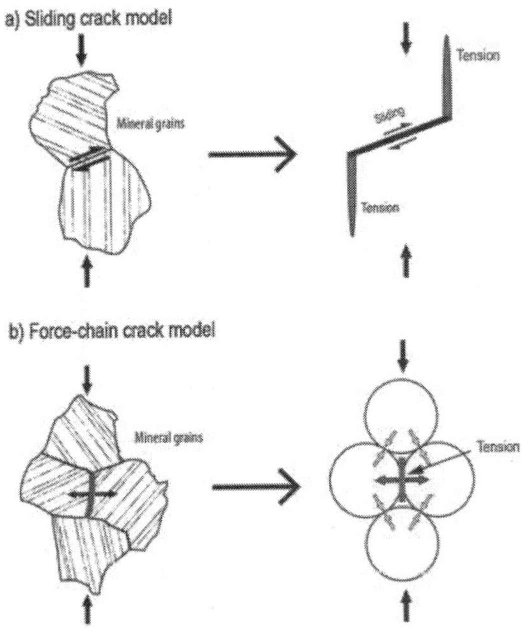

Figure 14. Two models commonly used to simulate cracking observed in heterogeneous assemblages of polygonal shaped minerals. Modified from Nicksiar and Martin (2013).

Potyondy and Cundall (2004) pointed out that systems composed of many simple objects commonly exhibit behavior that is much more complicated than that of the constituents. They listed the following characteristics that need to be considered in developing a rock mass model:

1) Continuously nonlinear stress–strain response, with ultimate yield, followed by softening or hardening.
2) Behavior that changes in character, according to stress state; for example, crack patterns quite different in tensile, unconfined- and confined-compressive regimes.

3) Memory of previous stress or strain excursions, in both magnitude and direction.
4) Dilatancy that depends on history, mean stress and initial state.
5) Hysteresis at all levels of cyclic loading/unloading.
6) Transition from brittle to ductile shear response as the mean stress is increased.
7) Dependence of incremental stiffness on mean stress and history.
8) Induced anisotropy of stiffness and strength with stress and strain path.
9) Nonlinear envelope of strength.
10) Spontaneous appearance of microcracks and localized macro fractures.
11) Spontaneous emission of acoustic energy.

Within the limitations of this document, it is clearly not feasible to present a summary of the many approaches that have been adopted in the numerical modeling of intact rock fracture initiation and propagation. Nor it is possible to judge the extent to which the requirements outlined above by Potyondy and Cundall (2004) have been met in these studies. A most useful DEM (discrete element method) approach is considered to be that given by Lan et al. (2010) who presented results of a study of fracture initiation and propagation in Äspö diorite and Lac du Bonnet granite. The mineral grain structures for these two crystalline rocks are shown in Fig. 15.

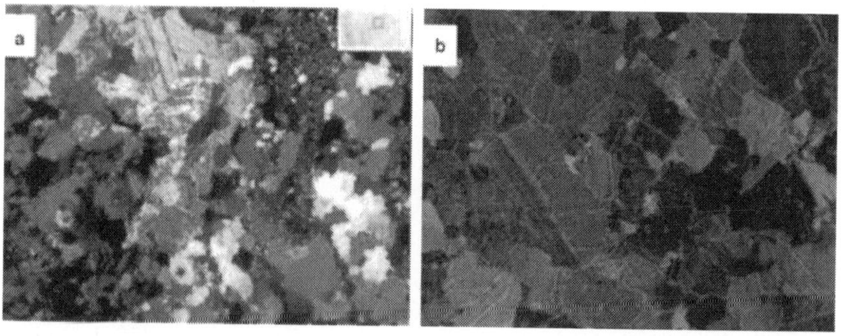

Figure 15. Example of mineral grain structure observed in polarized light thin section. (a) Äspö diorite. Width of the image is 4 mm (modified from Lampinen (2006)). (b) Lac du Bonnet granite. Combined polarized and fluorescent microscope image of specimen from Underground Research Laboratory in Canada. Width of image is 4 mm (modified from Åkesson (2008)). Images reproduced from Lan et al. (2010).

The program UDEC (Itasca, 2013) was used in this study and a Voronoi tessellation scheme was employed to create polygonal structures which closely simulated the mineral grain structures shown in Fig. 15. Each grain has a unique identity, location and material type and the average grain size distribution has also been simulated in these models. The properties of the principal grain minerals (plagioclase, K-feldspar and quartz with biotite in the Lac du Bonnet granite and with chlorite in the Äspö diorite) were exported to an ASCII file which was then imported into the UDEC model using the FISH internal macro-language. The model geometry is then created automatically in UDEC and the grains are made deformable by discretizing, each polygon using triangular zones. These deformable grains, which were unbreakable in the Lan et al. (2010) study, are then cemented together along their adjoining sides as shown in Fig. 16.

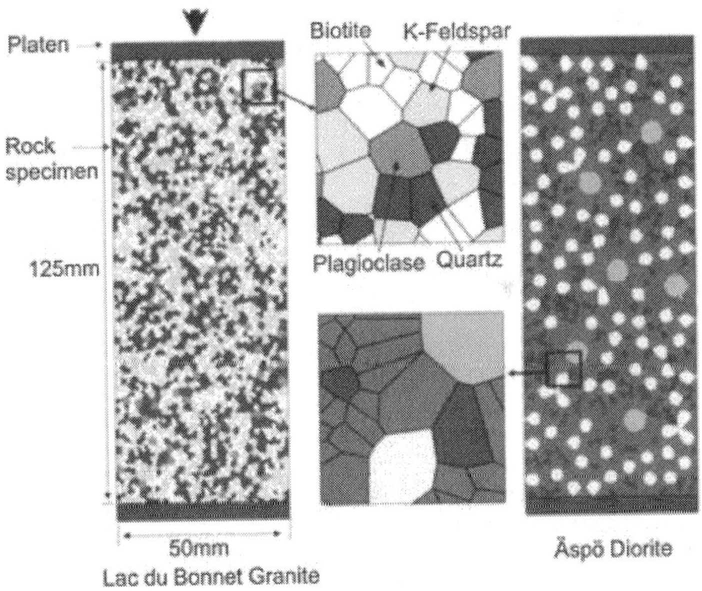

Figure 16. Layout for an unconfined compression test for a Lac du Bonnet granite sample and an Äspö diorite sample using the UDEC model. The different gray scales indicate the degree of mineral grain strength. Higher strength grains have a darker color. Reproduced from Lan et al. (2010).

Fig. 17 shows the results of two uniaxial compression tests carried out by Lan et al. (2010) on UDEC models of Äspö diorite and Lac du Bonnet granite. The stress–strain response, the crack initiation stress (σ_{ci}), the crack damage stress σ_{cd} and the peak stress σ_f show excellent agreement with those defined from laboratory tests on Äspö diorite by Staub and

Andersson (2004) and on Lac du Bonnet granite by Martin and Chandler (1994).

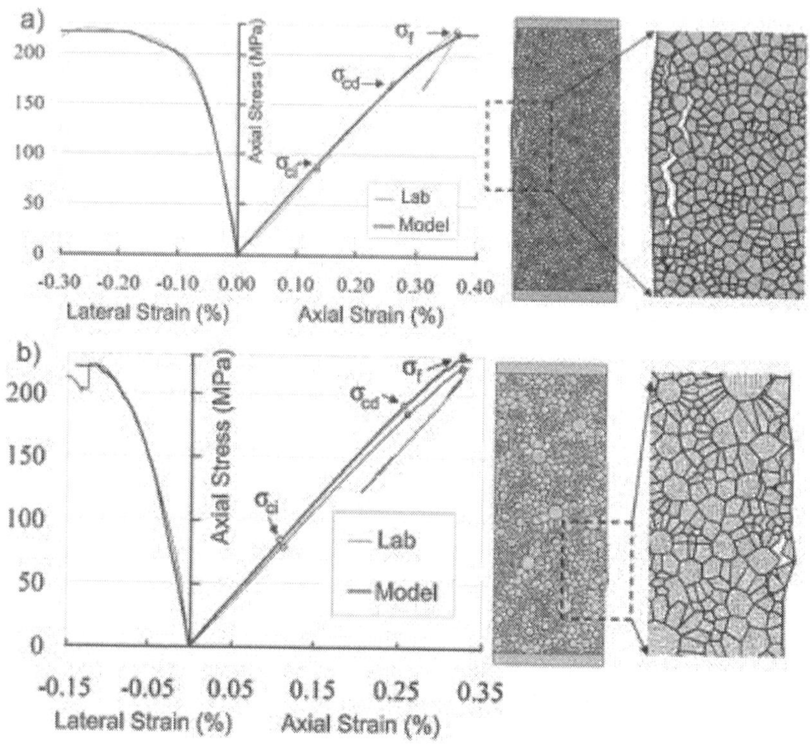

Figure 17. Calibrated stress–strain response with laboratory data for (a) Äspö diorite and (b) Lac du Bonnet granite. The drawings at the right show the damage pattern of the specimen. Images reproduced from Lan et al. (2010).

The paper by Lan et al. (2010) is a good example of the application of numerical modeling to the study of fracture initiation and propagation in intact rock and is recommended reading for anyone interested in this field. Much more work is required to bring this approach to maturity.

Professor E.T. Brown, in a foreword to the scoping study for the application of numerical methods to mass mining wrote: "In my opinion, the development of the bonded particle model based on the PFC and PFC3D distinct element codes by Dr. Peter Cundall and his co-workers at Itasca represents one of the most significant contributions made to modern rock mechanics research. It is now well established that this model has the ability to reproduce the essential, and some more subtle, features of the initiation and propagation of fracturing in rocks and rock masses." Numerical modeling has now progressed beyond the original bonded particle models developed by Itasca but Brown's comments remain valid as we look ahead to the research that remains to be done on these complex fundamental processes of failure initiation and propagation.

A word of warning. In the rush to get into print, many authors have published papers on numerical modeling in which pictures of fracture propagation in Brazilian disk tests or uniaxial compression tests have been included as a demonstration of the validity of the numerical approach used. It is relatively easy to produce results which appear to be credible for these tests but, unfortunately, in many cases the numerical methods used are immature and fail when applied to more complex problems.

In the following section, we briefly consider two approaches that may be used to evaluate fracture initiation and propagation in situ. The process is commonly referred to as spalling.

FRACTURE INITIATION AND PROPAGATION IN SITU: SPALLING

There are two practical issues associated with spalling: (1) identifying the conditions that will initiate spalling, and (2) defining the extent and depth of spalling. The results from two well-documented in situ experiments in crystalline rock are used to examine these practical issues.

Test tunnel of the URL Mine-by experiment

Martin et al. (1997) have described spalling observed in a test tunnel (Fig. 18) in massive Lac du Bonnet granite at a depth of 420 m below surface in the Underground Research Laboratory. The intent of the experiment was to study the damage resulting from stress redistribution associated with the full-face mining of 3.5 m diameter tunnel. The mining was carried out using line drilling and rock splitters to avoid the potential for blasting-induced damage. This technique allowed full-face 1-m advance increments.

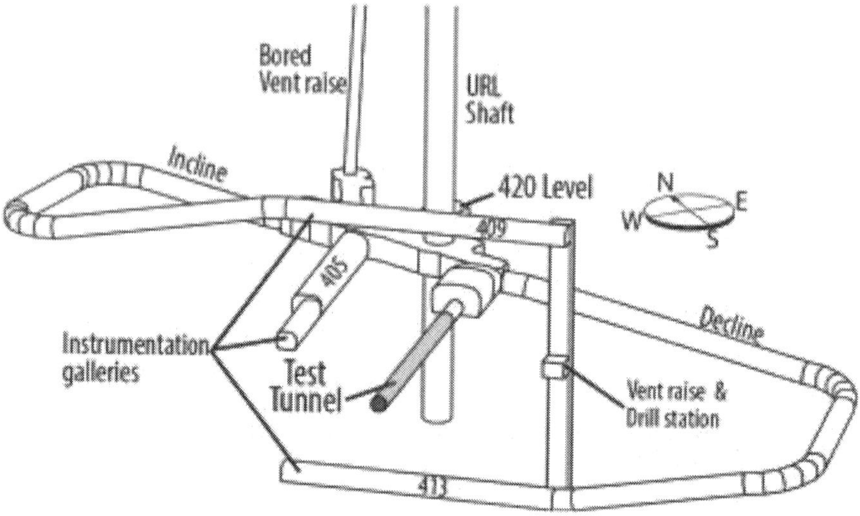

Figure 18. Layout of the Mine-by experiment at the 420 m level of the Underground Research Laboratory. Modified from Martin and Read (1996).

An extensive program of in situ stress measurements was carried out at this site and the rock mass surrounding the test tunnel. The in situ rock mass stresses at this location were a sub-vertical stress $\sigma_V = 11$ MPa and a sub-horizontal stress of $k\sigma_V = 60$ MPa, inclined at 11° to the horizontal, with an intermediate sub-horizontal stress of 44 MPa.

The spalling which occurred after excavation of the test tunnel is illustrated in Fig. 19. This spalling occurred as a relatively gentle fracture process during the excavation advance. The full extent of the spalling was only evident once the tunnel had been cleaned and loose spall remnants removed by scaling. When these remnants were removed in the floor, it was sufficient to trigger minor amounts of new spalling, suggesting the important role of small confining stress in controlling the spalling process.

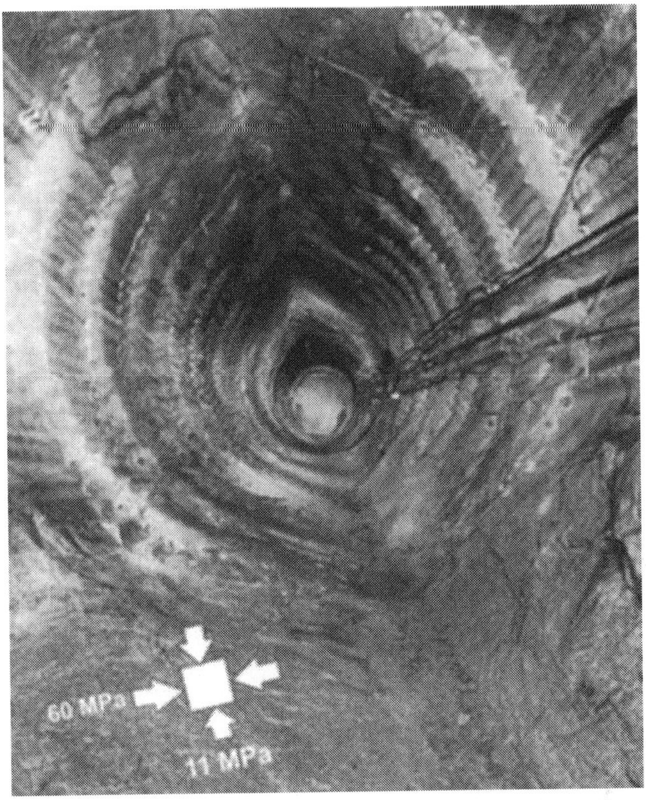

Figure 19. Spalling in the roof and floor of a circular test tunnel in the Underground Research Laboratory at Pinawa in Manitoba, Canada. Photo courtesy of AECL.

Fig. 20 shows the zone of potential spalling for the Lac du Bonnet granite in which the test tunnel was mined as well as the stress changes associated with the tunnel excavation. The measured in situ stresses, denoted by point *A* in Fig. 20, are equal to the principal stresses $\sigma_1 = 60$ MPa and $\sigma_3 = 11$ MPa in the rock before the tunnel was mined. The 11° inclination of the stress field can be ignored in the discussion which follows. After excavation, the minor principal stress in the tunnel wall is reduced to $\sigma_3 = 0$. The maximum principal stress on the tunnel roof and floor is given by $\sigma_{max} = 3\sigma_v(k-1) = 169$ MPa. These stresses are plotted as point *B* in Fig. 20.

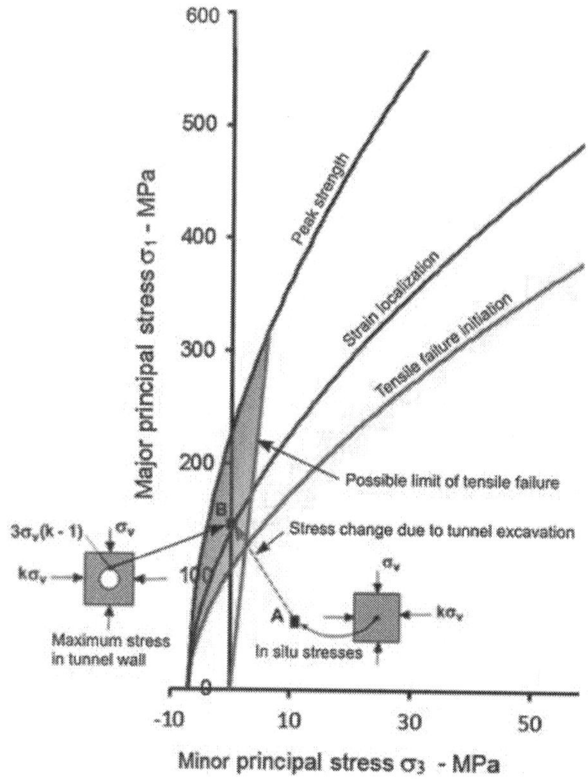

Figure 20. Definition of the zone of potential spalling in massive Lac du Bonnet granite, from Fig. 13. The changes in the stresses in the rock surrounding the tunnel are also shown as points *A* and *B* in this plot.

These roof and floor stresses fall just above the curve defining strain localization and well above the tensile failure initiation curve. Martin and Christiansson (2009) concluded that in the absence of in situ results, the laboratory crack initiation stress (see Fig. 13) could be taken as a lower bound for the spalling strength. More recently, Nicksiar and Martin (2013) compiled the crack initiation stress for a range of rock types. The results are illustrated in Fig. 21 and demonstrate the consistency of tensile crack initiation observed in laboratory tests. Using a spalling initiation criterion based on this approach is a useful first step and supported by recent experience. Diederichs et al. (2010) implemented this approach in a continuum model with good success.

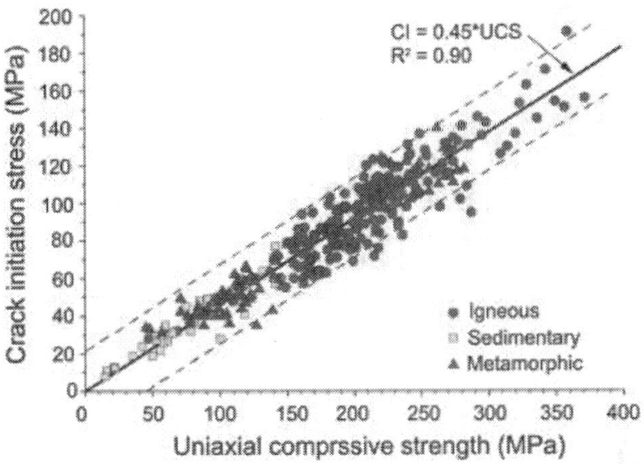

Figure 21. Tensile crack initiation in various rocks. Modified from Nicksiar and Martin (2013).

Knowing that spalling may occur, the next step is to establish the severity of the failure. The depth of the notch created by spalling is dependent upon the ratio of the maximum boundary stress to the uniaxial compressive strength (peak) or σ_{max}/σ_c as shown inFig. 22 (Martin et al., 1999). In the case of the test tunnel under discussion here, this ratio is 169/227 = 0.74 and hence the notch depth is approximately 0.3–0.4 times the tunnel radius according to Fig. 22.

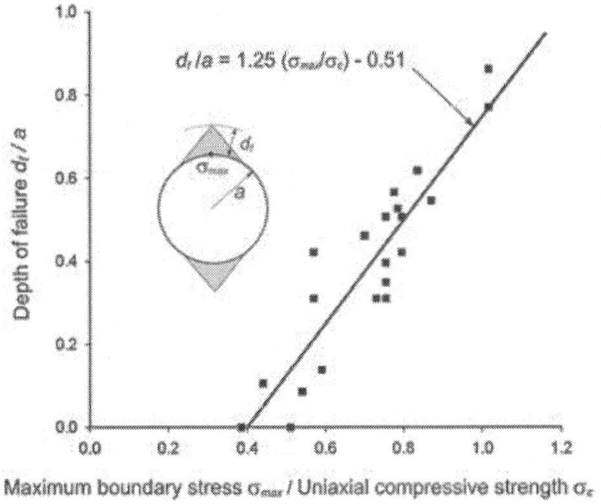

Figure 22. Observed spalling notch depths plotted against the ratio of maximum boundary stress to uniaxial compressive strength (Martin et al., 1999).

Experience with the application of the trend line in Fig. 22 indicates that the uniaxial compressive strength should be the mean uniaxial compressive strength value (Rojat et al., 2009). Determining the mean uniaxial compressive strength may appear straightforward. The scatter in the values for 13 samples of Lac du Bonnet granite is shown in Fig. 23. Notice that the mean value of 211 MPa is less than the σ_c value of 227 MPa given in Table 2 using the Hoek–Brown equations. This is a typical finding and in this case reflects the effect of the microcracks common in Lac du Bonnet granite. Despite the suggestions in the ISRM Suggested Methods, conducting uniaxial tests requires significant care in order to reduce the effect of uneven and/or misaligned sample ends. Simply conducting a large number of tests is no substitute for properly preparing the samples, and in many cases can be misleading as the mean value often decreases as the number of samples increases, due to poor quality control with sample preparation.

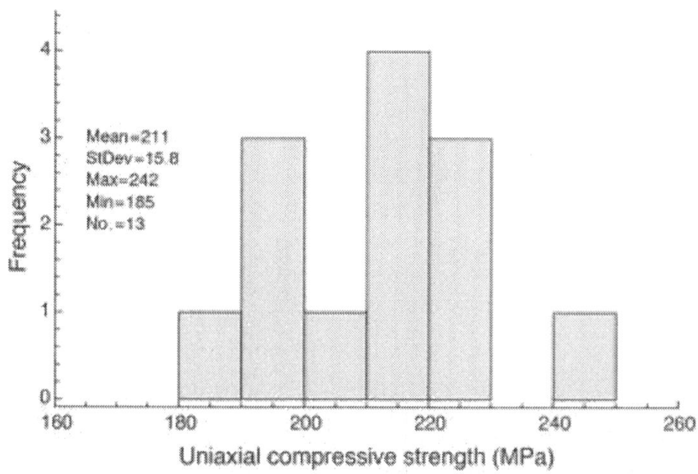

Figure 23. Example of the distribution of uniaxial compressive strength for 13 samples of Lac du Bonnet granite.

Fig. 24 and Fig. 25 show examples of mild spalling and severe rockbursting in underground excavations, representing the extremes of the process under discussion here. Practical experience suggests that shallow spalls are generally associated with pure tensile failure which causes thin slivers or plates of rock to peel off the tunnel surface. These occur with little "popping" and, once the maximum depth of the spall has been achieved, they remain stable provided that there are no changes in the surrounding stress field due, for example, to excavation of adjacent openings.

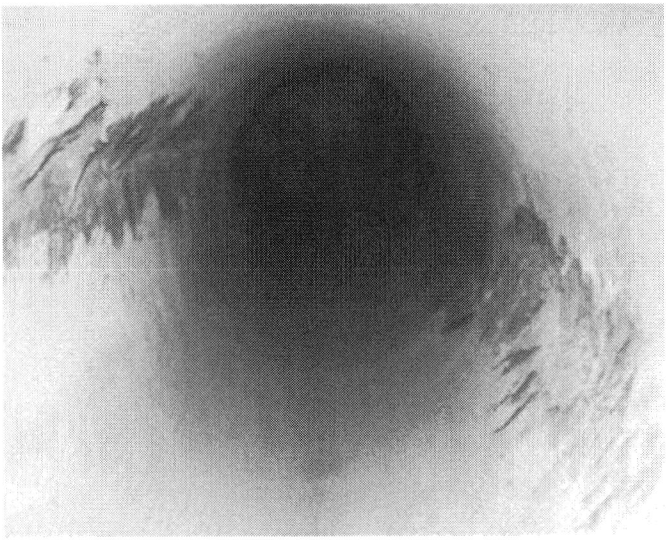

Figure 24. Mild spalling in the sidewalls of a vertical raise bored shaft in an underground mine.

Figure 25. Severe rockbursting in an access tunnel in a deep level gold mine in South Africa.

Deeper spalls, such as that in the Mine-by tunnel described above, are somewhat more complicated in that shear failure probably becomes involved as the notch tip moves away from the excavation boundary. Numerical analyses of this failure process have proved to be extremely challenging and it has to be said that much work remains to be done before the complex interaction of tensile and shear processes associated with deep spall notches can be predicted with any degree of confidence.

Rockbursts, such as that illustrated in Fig. 25, are probably associated with conditions in which the maximum induced boundary stress approaches the uniaxial compressive strength of the surrounding massive rock. These events involve implosion of the rock into the tunnel with the release of significant amounts of energy. The authors are not aware of any currently available numerical tools that offer any credible means of explaining or predicting the rockburst process.

Numerical simulation of spalling

The approach described above to estimate the onset and depth of spalling is a reasonable first step. But to make progress in the support interaction needed for tunnel design, proper numerical approaches are needed. While this is still an on-going research topic, one approach that goes from the laboratory calibration to spall prediction is described briefly below.

The nuclear industry in Finland and Sweden is preparing for the construction of a geological repository for used nuclear fuel at a depth of about 450 m. Their concept requires excavation of 1.75-m-diameter 8-m-deep boreholes and spalling is a design issue that must be addressed. Andersson et al. (2009) described and reported the results of a full-scale experiment (APSE) that examined the development of spalling around two of the large diameter boreholes. The APSE experiment was carried out at the 450-m level of the Äspö Hard Rock Laboratory in southern Sweden. Lan et al. (2013) described how the UDEC modeling work originally described by Lan et al. (2010) was used to model the spalling process. Fig. 26 shows the model configuration and the grain-scale geometry. The approach and the properties of the grains and the contacts were exactly the same as that given in Lan et al. (2010).

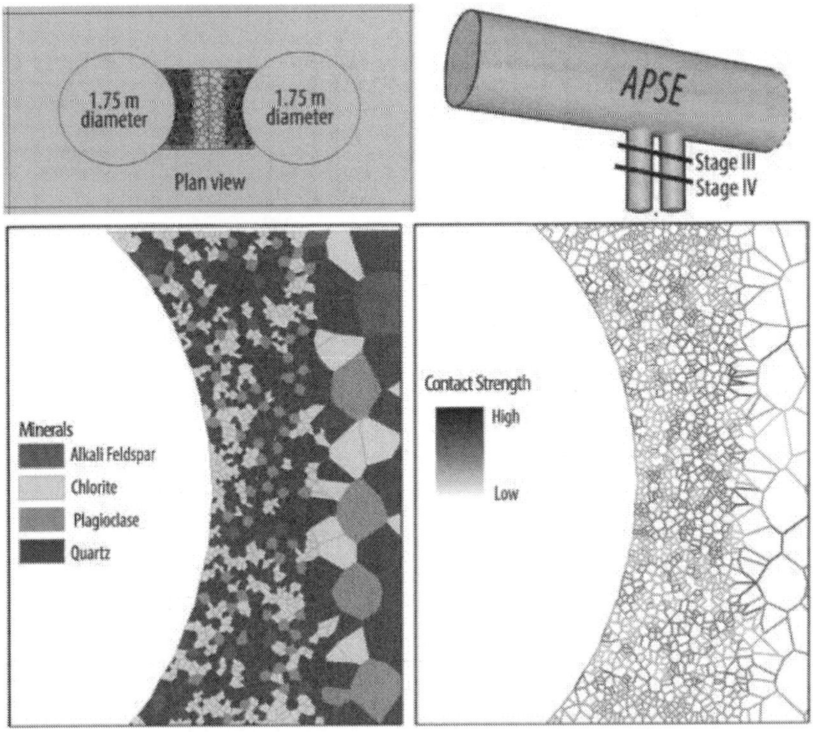

Figure 26. Grain-based UDEC model developed by Lan et al. (2010) and used to simulate the spalling process observed by Andersson et al. (2009). Modified from Lan et al. (2013).

The APSE experiment was unique because the magnitude of the stresses on the boundary of the large boreholes was controlled by excavation-induced and thermally induced stresses. The experiment demonstrated that in situ experiments could follow the same loading conditions and control that are normally associated with laboratory tests.

The configuration of the experiment allowed the boundary stresses to be applied gradually. This facilitated observing the spalling process at different stages. Fig. 27provides comparison of the results from the UDEC model with the visual observations at two loading stages. An important conclusion from the APSE work is that the findings related to fracture initiation and propagation that were observed in the Mine-by test tunnel in massive un-fractured granite were applicable to the fractured water-bearing rock mass of the APSE experiment.

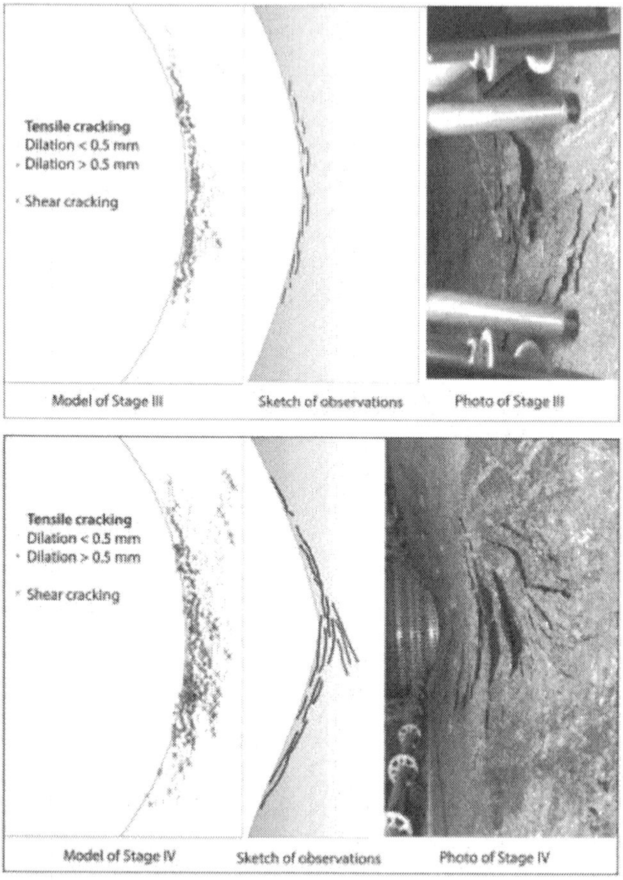

Figure 27. Modeled damage at stages III and IV compared with observation. Modeling result shows the distribution of tensile cracking and shear cracking at different damage stages.

The photograph and illustration are modified from Lan et al. (2013).

The approach described by Lan et al., 2010 and Lan et al., 2013 demonstrated that the properties of the laboratory tests can be used to evaluate the in situ spalling process when coupled with numerical approaches that capture all stages of brittle failure, i.e. from fracture initiation through to fracture propagation. While this approach holds much promise, it is still limited to two dimensions, and much work needs to be done before this approach becomes state of practice.

CONCLUSIONS

Our understanding of initiation and propagation of fracturing in intact rock has resulted from detailed analysis of the stress–strain data from laboratory-scale samples with dimensions in the range of 50 mm diameter. At low confining pressures, tensile fracturing initiates in these samples at 40%–60% of the uniaxial compressive strength and as loading continues, these tensile fractures increase in density and ultimately coalesce, leading to strain localization and macro-scale shear failure of the sample. The Griffith theory of brittle failure provides a simplified model assuming that all fractures initiate from the tips of inclined flaws, namely grain boundaries in the sliding-crack model. The Hoek–Brown failure envelope is used to capture the collapse load associated with this localized macro-scale shear fracture. However, it has been necessary to add a tension cutoff, based on a generalized fracture theory proposed by Fairhurst (1964) in order to accommodate the tensile failure observed in detailed laboratory tests.

Fracture initiation while tensile in nature is more difficult to be modeled. With the improvements in computing power, the discrete element codes have shown that the force-chain crack model is a viable alternative to explain the tensile fracture initiation coalition of tensile cracks and the final shearing of the specimens at higher confining stresses. Grain-based numerical models, based on the discrete element formulation, in which the grain size distributions as well as the physical properties of the component grains of the rock are incorporated, have proved to be very useful in studying these complex processes. They have also demonstrated that the approach based on laboratory processes is useful for capturing spalling, the in situ process of fracture initiation and coalescence. While this approach holds much promise, the current grain based models are still limited to two dimensions, and much research needs to be carried out before this approach becomes state of practice.

ACKNOWLEDGMENTS

The contribution of Dr. Connor Langford in deriving the solution to the generalized fracture criterion proposed by Fairhurst (1964) is gratefully acknowledged. This derivation is presented in the Appendix.

Professor Ted Brown reviewed the final manuscript and offered a number of valuable comments, which are also acknowledged.

APPENDIX

According to the generalized Fairhurst criterion:

1) If $w(w-2)\,\sigma_3 + \sigma_1 \leq 0$, failure occurs when $\sigma_3 = \sigma_t$;

2) If $w(w-2)\,\sigma_3 + \sigma_1 \leq 0$, failure occurs when

$$\frac{(\sigma_1-\sigma_3)^2}{(\sigma_1+\sigma_3)} = -2\sigma_t(w-1)^2\left\{1+\frac{2\sigma_t}{\sigma_1+\sigma_3}\left[\left(\frac{w-1}{2}\right)^2-1\right]\right\} \tag{A1}$$

where

$$w = \sqrt{\frac{\sigma_c}{|\sigma_t|}+1} \tag{A2}$$

Let $A = 2(w-1)^2$ and $B = [(w-1)/2]^2 - 1$, and rearrange Fairhurst's equation:

$$\sigma_1^2 + \sigma_1(A\sigma_t - 2\sigma_3) + \left(\sigma_3^2 + A\sigma_t\sigma_3 + 2AB\sigma_t^2\right) = 0 \tag{A3}$$

σ_1 can be written by

$$\sigma_1 = \frac{(2\sigma_3 - A\sigma_t) + \sqrt{(A\sigma_t - 2\sigma_3)^2 - 4\left(\sigma_3^2 + A\sigma_t\sigma_3 + 2AB\sigma_t^2\right)}}{2} \tag{A4}$$

REFERENCES

1. Åkesson U. Characterization of micro cracks using core disking. SKB Report P-08- 103. Stockholm: Swedish Nuclear Fuel and Waste Manage Co.; 2008. p. 43.
2. Andersson J, Martin CD, Stille H. The Äspö pillar stability experiment: Part II-rock mass response to coupled excavation-induced and thermal-induced stresses. International Journal of Rock Mechanics & Mining Sciences 2009;46(5):879e95.
3. Andriev GE. Brittle failure of rock materials. Rotterdam: A. A. Balkema; 1995.

4. Ashby MF, Hallam D. The failure of brittle solids containing small cracks under compressive stress. Acta Metallurgica 1986;34(3):497e510.
5. Bobich JK. Experimental analysis of the extension to shear fracture transition in Berea sandstone. MS Thesis. Texas, USA: Texas A&M University; 2005.
6. Brace WF. Brittle fracture of rocks. In: Judd WR, editor. State of stress in the earth's crust. New York: American Elsevier; 1964. pp. 111e80.
7. Cai M, Kaiser PK, Martin CD. A tensile model for the interpretation of microseismic events near underground openings. Pure and Applied Geophysics 1998;153: 67e92.
8. Cho N, Martin CD, Sego DS. A clumped particle model for rock. International Journal of Rock Mechanics and Mining Sciences 2007;44(7):997e1010.
9. Cundall PA, Strack O. A discrete numerical model for granular assemblies. Geotechnique 1979;29(1):47e65.
10. Cundall PA, Pierce ME, Mas Ivars D. Quantifying the size effect of rock mass strength. In: Potvin Y, Carter J, Dyskin A, Jeffrey R, editors. Proceedings of the1st southern hemisphere international rock mechanics symposium. Nedlands, Western Australia: Australian Centre for Geomechanics; 2008. pp. 3e15.
11. Diederichs MS, Carter T, Martin CD. Practical rock spall prediction in tunnels. In: Proceedings of ITA World Tunnel Congress; 2010. pp. 1e8. Vancouver, (CDROM).
12. Diederichs MS. Instability of hard rockmasses: the role of tensile damage and relaxation. PhD Thesis. Waterloo, Canada: University of Waterloo; 1999.
13. Diederichs MS. Rock fracture and collapse under low confinement conditions. Rock Mechanics and Rock Engineering 2003;36(5):339e81.
14. Fairhurst C. On the validity of the "Brazilian" test for brittle materials. International Journal of Rock Mechanics and Mining Sciences & Geomechanics Abstracts 1964;1(4):535e46.
15. Fairhurst C. Fundamental considerations relating to the strength of rock. In: Colloquium on rock fracture. Bochum, Germany: Ruhr Universitat; 1972. Updated in 2004 and available from http://www.itascacg.com. Germanovich JN, Dyskin AV. A model of brittle failure for material with cracks in uniaxial loading. Mechanics of Solids 1988;23(2):111e23.
16. Griffith AA. The phenomena of rupture and flow in solids. The Philosophical Transactions of the Royal Society London (Series A) 1921;221:163e98.
17. Griffith AA. Theory of rupture. In: Proceedings of the 1st international Congress of applied mechanics. Delft: Tech. Boekhandel en Drukkerij J Walter Jr; 1924. pp. 55e63.
18. Hoek E. Rock fracture under static stress conditions. Pretoria, South Africa: Council for Scientific and Industrial Research Report MEG 383; 1965.

19. Hoek E, Bieniawski ZT. Brittle fracture propagation in rock under compression. International Journal Fracture Mechanics 1965;1(3):137e55.
20. Hoek E, Brown ET. Empirical strength criterion for rock masses. Journal of the Geotechnical Engineering Division, ASCE 1980;106(9):1013e35.
21. Hoek E. The strength of jointed rock masses. Geotechnique 1983;33(3):187e223.
22. Itasca. UDEC 5.00 software reference manual. Minneapolis: Itasca; 2013.
23. Kemeny JM, Cook NGW. Crack models for the failure of rock under compression. In: Desai CS, Krempl E, Kiousis PD, Kundu T, editors. Proceedings of the 2nd international conference constitutive laws for engineering materials, theory and applications, vol. 1. London: Elsevier Science Publishing Co; 1987. pp. 879e87.
24. Lampinen H. Äspö Pillar Stability Experiment. Detailed geological mapping of pillar blocks. SKB Report IPR-05-24. Stockholm, Sweden: Swedish Nuclear Fuel and Waste Manage, Co.; 2006. p. 67.
25. Lan H, Martin CD, Andersson JC. Evolution of in situ rock mass damage induced by mechanical-thermal loading. Rock Mechanics and Rock Engineering 2013;46(1):153e68.
26. Lan H, Martin CD, Hu B. Effect of heterogeneity of brittle rock on micromechanical extensile behavior during compression loading. Journal of Geophysical Research 2010;115:B01202. http://dx.doi.org/10.1029/2009JB006496.
27. Lau JSO, Gorski B. Uniaxial and triaxial compression tests on URL rock samples from boreholes 207-045-GC3 and 209-069-PH3. CANMET Divisional Report MRL 92 e 025 (TR); 1992. p. 46.
28. Lorig LJ. Using numbers from geology. In: Proceedings of the 11th Congress of the international society for rock mechanics. London: Taylor and Francis; 2007. pp. 1369e77.
29. Martin CD, Chandler NA. The progressive fracture of Lac du Bonnet granite. International Journal of Rock Mechanics and Mining Sciences & Geomechanics Abstracts 1994;31(6):643e59.
30. Martin CD, Read RS. AECL's Mine-by experiment: a test tunnel in brittle rock. In: Aubertin M, Hassani F, Mitri H, editors. Proceedings of the 2nd North American rock mechanics symposium, vol. 1. Rotterdam: A.A. Balkema; 1996. pp. 13e24.
31. Martin CD, Read RS, Martino JB. Observations of brittle failure around a circular test tunnel. International Journal of Rock Mechanics and Mining Sciences 1997;34(7):1065e73. Martin CD. The effect of cohesion loss and stress path on brittle rock strength. Canadian Geotechnical Journal 1997;34(5):698e725.

32. Martin CD, Kaiser PK, McCreath DR. Hoek-Brown parameters for predicting the depth of brittle failure around tunnels. Canadian Geotechnical Journal 1999;36(1):136e51.
33. Martin CD, Christiansson R. Estimating the potential for spalling around a deep nuclear waste repository in crystalline rock. International Journal of Rock Mechanics and Mining Sciences 2009;46(2):219e28.
34. McClintock FA, Walsh JB. Friction on Griffith cracks in rock under pressure. In: Proceedings of the 4th US national Congress on applied mechanics, vol. 2. New York: Am. Soc. Mech. Eng.; 1962. pp. 1015e21.
35. Murrell SAF. The strength of coal under tri-axial compression. In: Walton WH, editor. Mechanical properties of non-metallic brittle materials. London: Butterworth Scientific Publications; 1958. pp. 123e45.
36. Nicksiar M, Martin CD. Crack initiation stress in low porosity crystalline and sedimentary rocks. Engineering Geology 2013;154:64e76.
37. Paterson MS, Wong TF. Experimental rock deformation e the brittle field. 2nd ed. New York: Springer-Verlag; 2005.
38. Pierce M, Mas Ivars D, Cundall P, Potyondy D. A synthetic rock mass model for jointed rock. In: Eberhardt E, Stead D, Morrison T, editors. Proceedings of Canada-U.S. rock mechanics symposium, vol. 1. London: Taylor & Francis Group; 2007. pp. 341e9.
39. Potyondy DO, Cundall PA. A bonded particle model for rock. International Journal of Rock Mechanics and Mining Sciences 2004;41(8):1329e64.
40. Potyondy DO. The bonded-particle model as a tool for rock mechanics research and application: current trends and future directions. In: Proceedings of the 7th Asian rock mechanics symposium, ARMS7. Seoul: ARMS; 2012. pp. 1e33.
41. Ramsey J, Chester F. Hybrid fracture and the transition from extension fracture to shear fracture. Nature 2004;428:63e6.
42. Read RS, Martin CD. The underground research laboratory Mine-by experiment e a research perspective on tunnel design. Canadian Tunnelling 1991:75e88.
43. Rojat F, Labiouse V, Kaiser PK, Descoeudres F. Brittle rock failure in the Steg Lateral Adit of the Lötschberg base tunnel. Rock Mechanics and Rock Engineering 2009;42(2):341e59.
44. Scholtès L, Donzé FV. A DEM model for soft and hard rocks role of grain interlocking on strength. Journal of the Mechanics and Physics of Solids 2013;61(2):352e69.
45. Staub I, Andersson JC. Äspö pillar stability experiment geology and mechanical properties of the rock in TASQ. SKB Report R-04-0. Stockholm: SvenskaKärnbränslehantering AB; 2004.

46. Zuo JP, Li HT, Xie HP, Ju Y, Peng SP. A nonlinear strength criterion for rock-like materials based on fracture mechanics. International Journal of Rock Mechanics and Mining Sciences 2008;45(4):594e9.

CITATION

E. Hoek, C.D. Martin, Fracture initiation and propagation in intact rock – A review, Journal of Rock Mechanics and Geotechnical Engineering, Volume 6, Issue 4, August 2014, Pages 287-300, ISSN 1674-7755, http://dx.doi.org/10.1016/j.jrmge.2014.06.001.

Index

T

U